Elements of Meteorology

Elements of Meteorology

ALBERT MILLER
and
JACK C. THOMPSON

*Meteorology Department
San Jose State College*

CHARLES E. MERRILL PUBLISHING COMPANY
A Bell & Howell Company
Columbus, Ohio

Library of Congress Catalog Card Number: 78-98476
International Standard Book Number: 0-675-09554-9

PRINTED IN THE UNITED STATES OF AMERICA

3 4 5 6 7 8 9 10–74 73 72 71

To Rosa and Violette,

    without whose patience and understanding
this book could not have been written.

Preface

The popularity of the weather as a topic of daily conversation is certainly warranted: Like a fish in the ocean, man spends most of his life confined to a very shallow layer of the atmosphere. His physical and psychological state—indeed, his very life—depends on his atmospheric environment.

Aside from the use of its constituents in biological processes, the atmosphere controls life in many ways. It acts as an umbrella or shield, filtering various types of electromagnetic radiation and high-energy particles that come from the sun and space, and burning up most meteorites before they reach the earth's surface. Atmospheric winds transport heat and moisture and, in the process, produce mixing of the air, thus creating more uniform conditions on the earth than would otherwise exist. The same winds drive the ocean currents, produce waves, erode the soil, and transport pollen and insects. The weather that destroys man's structures and disrupts his systems of communication and transportation also provides the unsalted water that he needs for survival. The sounds man hears, the scents he smells, and the sights he sees are all affected by the state of the atmosphere. In fact, man and all other forms of life are so delicately tuned to their atmospheric environment that they can tolerate very little change.

The science of meteorology attempts to gain insight into the physical laws or relationships that determine the state of the atmosphere. The practical applications of such knowledge include: (1) prediction of future weather to guide the planning of man's activities; (2) adaptation of man's activities to the weather; and (3) modification of weather. Most people are familiar with the first application because of the daily weather forecast, but those whose activities are especially sensitive to the environment have long been concerned with the second also. The third application, weather modification, has received sporadic, generally minor attention until recently, when attempts to increase rainfall by cloud seeding have renewed interest among both laymen and professional meteorologists.

Prediction is a fundamental task of all sciences; yet, public weather forecasting, even after a hundred years, is still singled out as the subject of countless jokes. For a period of up to one or two days, the accuracy of weather forecasts is high—though certainly not perfect—but beyond a couple of days, the reliability falls off markedly. If the astronomer can forecast an eclipse years ahead without a miss, why can't the meteorologist, who has the same basic physical laws at his command, foretell exactly when tomorrow's rain will begin?

Hopefully, the answer to this question will become clear in this book. The laws of physics *can* be applied to the atmosphere to explain its state. But the "state" of the atmosphere, or "weather," is a composite of many elements which are distributed in an infinite number of patterns in space and time. The complexity of such patterns is so great that progress in imposing "order" on the atmosphere to predict its behavior has been painfully slow.

In the first two-thirds of this book, a discussion of atmospheric processes and phenomena, two basic concepts are employed: One is that the atmosphere is a giant *heat engine*, converting energy from one form to another. The other is that circulation systems exist in a wide range of sizes or *scales*, all of which are interdependent. The remainder of the book is devoted to the practical applications of meteorology: prediction, environmental adaptation, and weather modification.

A great deal of reference material has been put in the appendices. The problems at the end of each chapter also contain useful information. These appear in approximately the order of presentation of topics in each chapter; the instructor may wish to make a selection appropriate to the background of his students.

The text is meant for those who are curious about their physical environment but have little formal training in physics and mathematics.

Despite the limitation of the use of established physical principles
and mathematical manipulation, the authors have attempted a modern,
non-superficial explanation of the workings of the atmosphere without becoming so enmeshed in details as to lose the thread of the
"story" of meteorology. Where quantitative or mathematical material
has been considered desirable, an alternative descriptive explanation
also has been provided as an aid to qualitative understanding of the
basic concept.

The authors wish to express their appreciation to Newton A. Lieu-
rance, who made comments on the section on aviation; James R.
Ballard, who read the portion on agriculture, Francis D. Beers, who
provided suggestions for the section on forestry, and William Long,
who reviewed the entire manuscript.

Albert Miller
Jack Thompson

San Jose, California
January, 1970

Contents

1

The Structure
of the Atmosphere

1-1. Origin

Although every planet in our solar system, with the
possible exception of Mercury, has an atmosphere, the
earth's atmosphere is probably the only one that is
composed almost entirely of nitrogen and oxygen.
Mars and Venus have atmospheres composed mostly of
carbon dioxide, while Jupiter's and Saturn's atmospheres contain mostly hydrogen, helium, methane, and
ammonia. The origin of the earth's particular kind of
atmosphere is still the subject of much speculation.
However, one thing seems fairly certain: When the
earth was formed, about 4½ billion years ago, it was
probably much too hot to retain *any* atmosphere that

1

it may have had to begin with. This is because high temperature implies high molecular speeds. What keeps the moving gas molecules from escaping from a planet is gravitational attraction, which depends on the massiveness of the bodies. Thus, small, hot planets, such as Mercury, can hold only the most massive gas molecules. Earth is massive enough, yet cool enough at the *present* time, to retain its thin envelope of gases.

At the outer limits of the earth's atmosphere, there is, even today, a slow seepage of molecules to outer space. At levels above 600 km, the gases are thin enough and hot enough so that some of the lighter, faster molecules can escape the gravitational pull of the earth. Because the upward "escape velocity" required for a particle to escape from the earth is quite high—about 11.3 km/sec (compared to 2.4 km/sec for the much smaller moon), only the very light gases, such as hydrogen and helium, can leave the earth at a significant rate. It is estimated that the much heavier oxygen and nitrogen molecules of today's atmosphere would require about 10^{45} and 10^{51} years, respectively, to be completely lost through such seepage.

At the beginning of earth's existence, its original atmosphere was probably composed chiefly of a smelly, noxious mixture of methane (CH_4) and ammonia (NH_3), which are even now important constituents of Jupiter, Saturn, Uranus, and Neptune. It is fairly certain that the first atmosphere was devoid of free oxygen. According to one theory, the earth's present atmosphere did not evolve until much of the original one had been driven off and until after the earth had started to cool. This new atmosphere was created when gases that had been dissolved in the molten rock bubbled out of the surface. These first gases were probably mostly steam, with some carbon dioxide and nitrogen (the principle gaseous emissions from active volcanoes even today). As cooling continued, the water vapor condensed to form the great oceans. The liquid water gradually absorbed most of the atmosphere's carbon dioxide, thus leaving nitrogen as the predominant gas. The oxygen in the atmosphere is believed to have appeared only after the appearance of primitive plant life (some 800 million years ago) which could act on the carbon dioxide through photosynthesis to form oxygen.* The present mixture of essentially nitrogen and oxygen has therefore probably been in existence only since some time after the advent of photosynthesis.

---

* An alternate theory for the creation of the oxygen in the atmosphere centers on the photolysis of water vapor and the subsequent escape of hydrogen from the earth.

1-2. Present Composition

The thin shell of air that now surrounds the earth is a mechanical mixture of many gases in which are suspended, in quite variable amounts, particles of liquid and solid matter. Although nitrogen and oxygen are by far the most abundant constituents, there are other gases, as well as liquid and solid particles, that are of great significance.

TABLE 1-1

The Composition of the Lower Atmosphere

Permanent constituents		Variable constituents	
Constituent	*% by volume*	*Constituent*	*% by volume*
Nitrogen (N₂)	78.084	Water vapor (H₂O)	<4
Oxygen (O₂)	20.946		
Argon (A)	0.934	Ozone (O₃)	$<.07 \times 10^{-4}$
Carbon dioxide (CO₂)	0.033° (up to 0.1)	Sulfur dioxide (SO₂)	$<1 \times 10^{-4}$
Neon (Ne)	18.18×10^{-4}	Nitrogen dioxide (NO₂)	$<0.02 \times 10^{-4}$
Helium (He)	5.24×10^{-4}		
Krypton (Kr)	1.14×10^{-4}	Ammonia (NH₃)	Trace
Xenon (Xe)	0.087×10^{-4}	Carbon monoxide (CO)	$\sim 0.2 \times 10^{-4}$
Hydrogen (H₂)	0.5×10^{-4}		
Methane (CH₄)	2.0×10^{-4}	Dust (soot, soil, salts)	$<10^{-5}$
Nitrous oxide (N₂O)	0.5×10^{-4}	Water (liquid and solid)	<1
Radon (Rn)	6×10^{-18}		

° Global average.

A summary of what is now known about the composition of the atmosphere near sea level is presented in Table 1-1; some indication of how the composition changes with height is shown in Figure 1-1. Below about 80 km, the gases of the atmosphere are relatively well mixed. In this layer, known as the *homosphere*, the proportion of each constituent gas, with few exceptions, is fairly constant throughout. In contrast, in the *heterosphere*, above 80 km, the various gases have tended to stratify in accordance with their weights, as occurs with liquids of different densities.

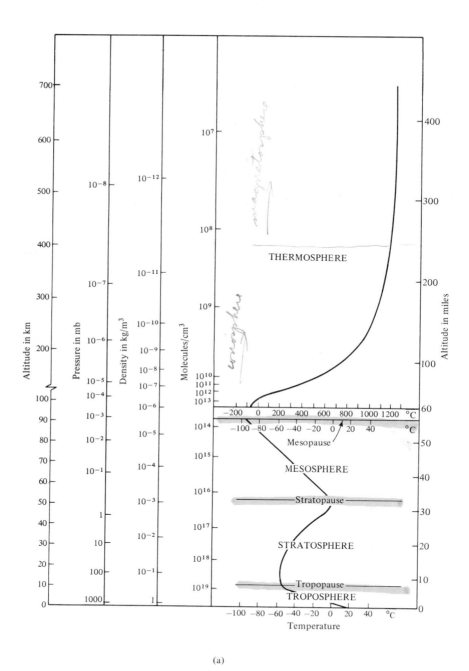

FIG. 1-1. Vertical distribution of atmospheric properties and phenomena.

4

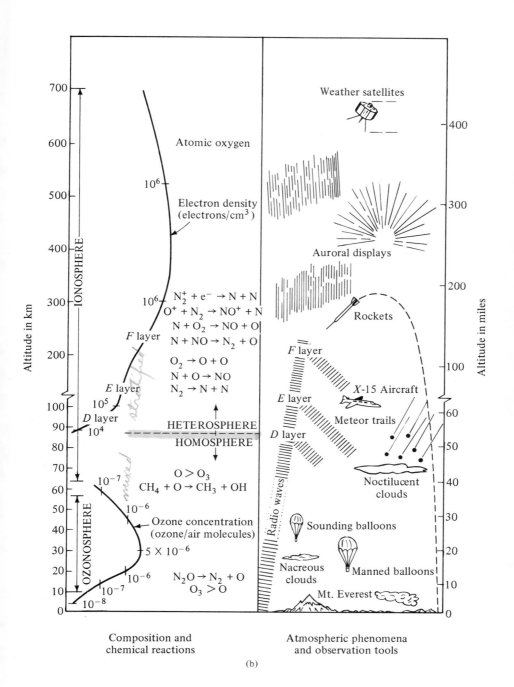

Altitude in km

Altitude in miles

IONOSPHERE

700
600
500
400
300
200
100
90
80
70
60
50
40
30
20
10
0

Atomic oxygen

10^6

Electron density
(electrons/cm^3)

10^6

F layer

E layer

10^5

D layer
10^4

$N_2^+ + e^- \rightarrow N + N$
$O^+ + N_2 \rightarrow NO^+ + N$
$N + O_2 \rightarrow NO + O$
$N + NO \rightarrow N_2 + O$

$O_2 \rightarrow O + O$
$N + O \rightarrow NO$
$N_2 \rightarrow N + N$

stratified

HETEROSPHERE

HOMOSPHERE

mixed

10^{-7}

$O > O_3$
$CH_4 + O \rightarrow CH_3 + OH$

OZONOSPHERE

10^{-6}

Ozone concentration
(ozone/air molecules)

5×10^{-6}

10^{-6}

$N_2O \rightarrow N_2 + O$
$O_3 > O$

10^{-7}

10^{-8}

Weather satellites

Auroral displays

Rockets

F layer

X-15 Aircraft

E layer

Meteor trails

D layer

Noctilucent
clouds

Radio waves

Sounding balloons

Nacreous
clouds

Manned balloons

Mt. Everest

400
300
200
100
60
50
40
30
20
10
0

Composition and
chemical reactions

Atmospheric phenomena
and observation tools

(b)

5

The Lower Atmosphere

Excluding the variable constituents, nitrogen and oxygen occupy about 99 per cent of the volume in the homosphere, argon comprises almost 1 per cent, and carbon dioxide typically occupies about 0.03 per cent, although it can reach as much as 0.1 per cent (Table 1-1). Even in their peak concentrations, the variable constituents (principally gaseous water) rarely occupy more than 4 per cent of the total volume. Some of the minor constituents are far more important in meteorological processes than are the more abundant nitrogen, oxygen, and argon. For example, carbon dioxide, water vapor, ozone, dust, and water in the liquid and solid forms strongly influence the radiant energy received from the sun and emitted by the earth. (The details of their effects on radiation will be covered later in this chapter and in Chapter 3.) Water plays such a variety of roles in weather production that we shall dedicate several sections of this book to it. The following paragraphs mention some significant characteristics of the more important "minor" constituents of the atmosphere.

The air always contains some water in the gaseous state and sometimes as much as 4 per cent by volume. Water is the only substance that can exist in all three states—gaseous, liquid, and solid—at the temperatures that exist normally on the earth. The cycle of transition between these states goes on continuously and plays an important role in maintaining life. But, in addition, these *phase changes* of water play another role in the atmosphere which is significant to the meteorologist: During the transition from a liquid or solid to a gaseous state, water molecules take up some heat energy which they obtain from the air in which they contained, and when they revert to the liquid or solid state they release the same amount of energy to their environment. Thus, heat consumed at one place during evaporation may be released at an entirely different place during condensation. This is an effective way of transporting heat over great distances. This "latent heat" is the source of energy for many of the atmosphere's storms, playing a dominant role in such phenomena as thunderstorms, tornadoes, and hurricanes.

Carbon dioxide is a natural constituent of the atmosphere. It is an efficient absorber of radiant energy in certain wavelengths of the infrared region of the spectrum (Chapter 3) and is, therefore, of considerable interest to the meteorologist who is concerned with the flow of energy into and out of the atmosphere. There is a natural cycling of this gas between the atmosphere and the ocean and the earth's surface as shown in Table 1-2.

TABLE 1-2

The Carbon Dioxide Cycle

Amount stored in:	billions of tons
1. Atmosphere	2,300
2. Oceans	130,000
3. Earth (mostly in fossil fuels)	40,000

Yearly flux	billions of tons per year
1. From biosphere to atmosphere (respiration and decay of organic material)	60
2. From atmosphere to biosphere (photosynthesis)	60
3. From oceans to atmosphere	100
4. From atmosphere to oceans	100
5. Into atmosphere from earth's interior (volcanoes, hot springs)	0.1
6. Into new fossils within earth	<0.1
7. Into atmosphere through man's combustion of fuels	6
8. Into atmosphere from newly cultivated soil	2

Note the relatively small amounts going into and out of the atmosphere each year compared to the amounts stored in each. However, there is speculation that man may be significantly changing the amount of carbon dioxide in the air through his increased burning of fossil fuels during the past half-century or so,* and that this may be affecting the atmosphere's heat balance. There is evidence that the amount of carbon dioxide in the vicinity of industrialized urban areas is significantly higher than the global average (Table 1-1), and the warmth of cities may be partly due to this fact.

Ozone is found in very minute quantities near the surface of the earth, usually comprising less than one part in a hundred million. If all the ozone in the atmosphere could be brought down to sea level pres-

---

* Carbon dioxide emissions from the combustion of coal, oil, and gas in the U.S. are estimated to have increased ninefold between 1890 and 1965—from 390 million to 3540 million tons per year. By the year 2000, the rate of CO_2 emission is expected to reach 9325 million tons per year.

sure and temperature, it would form a layer only about 2.5 mm thick. However, although the concentration of ozone is low at all levels of the atmosphere, there is a sharp peak near the altitude of 25 km, as can be seen from the curve of Figure 1-1. Despite the small quantities, ozone is quite significant in the radiant energy transfer that goes on in the atmosphere. Because of its strong absorption of ultraviolet light from the sun, very little of these lethal wavelengths arrive at the surface of the earth. The ozone (O_3) of the atmosphere is believed to form when an atom of oxygen (O), a molecule of oxygen (O_2), and a third "catalytic" particle, such as nitrogen, collide. The atomic oxygen is formed in the atmosphere by the splitting of molecular oxygen under the action of very short waves of solar radiation. (Note from Figure 1-1 that the production of atomic oxygen becomes especially prominent in the heterosphere.) The maximum of ozone near 25 km is apparently due to a balance of two factors: the availability of very short wavelengths of solar energy needed to produce atomic oxygen, which is gradually depleted as it traverses the upper layers of the atmosphere, and a sufficient density of particles to bring about the multiple collisions required.

Fairly high concentrations of ozone often occur in the lowest few hundred meters of the atmosphere, especially over urban areas. Ozone, which is a corrosive, toxic gas, is an important constituent of the so-called "photochemical smog" that afflicts many large cities. The atomic oxygen required for the reaction described above is formed in smog principally through the action of solar radiation on nitrogen dioxide, a product of combustion.

A great variety of solid particles are suspended in the air. These include fine dust swept up by the wind from exposed soils; soot from grass and forest fires, industrial fires, industrial plants, and volcanoes; pollen and microorganisms lifted by the wind; meteoritic dust; and salts injected into the atmosphere from the ocean's surface. Large particles are too heavy to remain long in the air, but there are many, so small that they cannot be seen individually with the naked eye, that remain suspended for months and even years. Although most particles remain in the atmosphere for less than a few days, during that time they can be transported great distances by the wind. For example, spores and pollen have been observed to travel more than 1000 miles. Fine sand from the Sahara Desert has been observed to fall in rain over the Swiss Alps, sometimes giving the glaciers a pink tinge. Sea salt (mostly NaCl) can be carried thousands of miles inland from the ocean. Sea salt gets into the air principally through the evaporation of minute droplets that have been injected into the air when the myriads of air bubbles that are created in turbulent ocean water burst as they hit the surface.

Since the source of most solid particles in the air is the earth's surface, their number normally decreases very rapidly with height above the ground. Typically, the concentration at 5 or 6 km is only a few per cent of that near the surface. The only significant extraterrestrial source of material is meteoritic dust, of which some 5 million tons enter the atmosphere each year.

Occasionally, particles from the earth's surface are forced to high levels of the atmosphere by volcanic eruptions or nuclear explosions. For example, minute particles of dust and ash were thrown high into the atmosphere by the violent eruption of the volcano Krakatoa in the East Indies in August of 1883; these particles were observed to circle the globe for at least two years. Nuclear explosions at the surface throw thousands of tons of vaporized soil into the air which later condense. Although most of this material is usually confined to the lowest 10 km of the atmosphere and falls out or is "washed out" of the air by rain within a couple of days, some small particles of radioactive debris reach into the dry stratosphere (see p. 16), where they may remain for as long as five years. Since certain radioisotopes, like strontium 90 and cesium 137, have long half-lives (27.7 and 30.5 years respectively), the stratosphere serves as a reservoir for this man-made radioactive material.

The number and sizes of non-gaseous particles of all types in the atmosphere are extremely variable as can be seen from values given in Table 1-3. Near the ground, the concentration of all particles can range from only one or two hundred per cubic centimeter in very clean air to several million per cubic centimeter in smoke-laden air. The average over cities is about 150,000/cm^3.

Man's activities add a great deal of dust to his atmospheric environment. To give some idea of how much, Table 1-4 presents the average mass of solid particles found in the air in certain cities of the United States during one year.

Fortunately, most particles are filtered out by the nose and throat before reaching the lungs, many are exhaled by the lungs, and some are expectorated. Atmospheric dust may also affect the climate by preventing some of the sun's radiation from reaching the earth's surface. It also reduces visibility by "dispersing" the rays of light before they reach the eye.

A large proportion of the other minor variable gases in the atmosphere (Table 1-1), such as sulphur dioxide, nitrogen dioxide, and car-

TABLE 1-3

Characteristic Sizes and Concentrations of Atmospheric Constituents (Near Sea Level)

Type	Diameter° (mm) Range of sizes	Typical	Concentration (no./cm³) Range	Typical	Approx. terminal velocity (cm/sec)
Gas molecules		1.9×10^{-7}		2.5×10^{19}	
Small ions	$(0.15–1) \times 10^{-5}$		$(1–7) \times 10^{2}$		
Large ions	$(1–20) \times 10^{-5}$		$(2–20) \times 10^{3}$		
Small (Aitken)† condensation nuclei‡	$(0.1–4) \times 10^{-4}$		$10–10^{5}$	10^{3}	$10^{-5}–10^{-3}$
Large nuclei	$(4–20) \times 10^{-4}$		$1–10^{3}$	10^{2}	$10^{-3}–7 \times 10^{-2}$
Giant nuclei	$(20–1000) \times 10^{-4}$		$10^{-4}–10$	1	$7 \times 10^{-2}–7 \times 10^{-1}$
Dry haze	$(1–100) \times 10^{-4}$		$10^{3}–10^{5}$		$10^{-3}–10^{-2}$
Fog and cloud droplets	$(1–200) \times 10^{-3}$	20×10^{-3}	$25–600$	300	$.01–70$
Drizzle	$(2–40) \times 10^{-2}$	30×10^{-2}	$1–10$		$1–170$
Raindrops	$0.4–4$	1	$10^{-3}–1$		$170–900$
Snow crystals	$0.5–5$	2	<10		$30–100$
Snowflakes	$4–20$	10	$10^{-3}–1$		$80–200$
Hail	$5–75+$ (Largest: 140)	15	$10^{-6}–10^{-1}$		$800–3500+$

° The term "diameter" is used loosely here, since most particles, including gas molecules, are not perfect spheres. For crystals and snowflakes, the figures given are the maximum dimension.

† Named for the British physicist who studied particles that serve as nuclei for condensation (Chapter 2).

‡ About half of these small nuclei carry a net electrical charge and are, therefore, also included in the number of large ions in the atmosphere.

TABLE 1-4

Particulate Concentrations of Some U.S. Cities (Means for 1958)
(grams $\times 10^{-4}/m^{3}$)

Louisville	228	Cincinnati	143
Los Angeles	213	Denver	110
Pittsburgh	167	San Diego	93
New York	164	San Francisco	80

bon monoxide, is introduced by man. These gases are usually by-products of some form of combustion and are hazardous to health.

The Upper Atmosphere

The major constituents of the atmosphere remain virtually unchanged up to 80 or 90 km, although there are significant variations in such minor constituents as ozone, dust, and water vapor. But above this level (at which point only 2/1,000,000 of the total atmosphere remains), the relative amounts and types of gases are no longer constant with height. Within this upper zone of thin air, known as the *heterosphere*, very short waves of ultraviolet and X-ray radiation from the sun cause various "photochemical" effects on the gases, such as the splitting of O_2 mentioned in the discussion of the formation of ozone. A photochemical effect or reaction is one in which the structure of a particle is changed when it absorbs radiant energy. A molecule may gain so much energy that it splits into two parts. Or, a molecule or atom may have one or more of its electrons stripped from it. Such photochemical reactions occur on a large scale in the heterosphere. Molecular oxygen is split into two atoms under the action of ultraviolet light at altitudes as low as 50 km; the amount of atomic oxygen produced in this fashion increases with height until the concentration begins to approach that of O_2 and, at 200 km, atomic oxygen is even more abundant than molecular nitrogen.

Molecular nitrogen is relatively immune to dissociation by ultraviolet radiation, but nitrogen, as well as other atmospheric gases can be readily *ionized* by very short waves of energy. (Particles are said to be *ionized* when electrons have been ejected from their atoms.) A large part of the air in the heterosphere consists of positively charged particles and free electrons. (In Figure 1-1, compare the number of electrons per cubic centimeter with the number of molecules per cubic centimeter in the heterosphere.) The rate of ion production depends on two factors: the density of molecules and atoms available for ionization and the intensity of the impinging radiation; the first will decrease, while the second increases with increasing height above the earth's surface. The capture of electrons by positively charged ions also depends on the density of the air, since the likelihood of collisions is greater if the particles are crowded than if they have a lot of room. As a result of a combination of these factors that determine ion production and destruction rates, there is a maximum of free electrons at about 350 km, although the distribution is rather complicated, forming into rough layers (designated as *D*, *E*, and *F* in Figure 1-1).

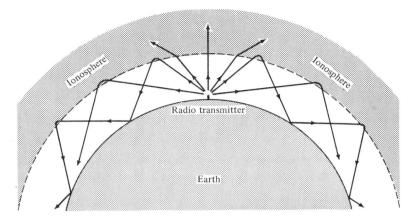

FIG. 1-2. Reflection of radio waves of ionosphere. Near-vertical rays
pass through the ionosphere, but those that strike at a critical angle (depending on frequency) are reflected. The earth's
surface, which is also a conductor, reflects some of the rays
skyward.

This electrically charged portion of the atmosphere, known as the
ionosphere, is very useful for radio communications, since it reflects
radio waves. Around-the-world transmissions are accomplished by
bouncing radio waves, which move in straight lines, between the
ionosphere and the earth's surface (Figure 1-2).

The distribution of electron density with height is not constant.
There is a diurnal variation in the strength of some layers due to
changes in the intensity of solar radiation. In addition, there are occasional *sudden ionospheric disturbances* (S.I.D.) and "ionospheric
storms" that are associated with disturbances on the sun. The S.I.D.
lasts for 15 to 30 minutes and is produced by bursts of ultraviolet
energy from the sun that cause a sudden increase in the production of
electrons. Since electrons absorb part of the radio energy that strikes
them, a sudden increase in their number may actually smother the
radio energy, leading to "fadeouts" of communications on the sunlit
side of the earth. Ionospheric storms, which can occur during the day
or night and last for hours or even days, are believed to be caused by
a stream of charged particles emitted from the sun. These fast-moving
particles, guided toward the magnetic poles* by the earth's magnetic

---

* The magnetic poles do not coincide with the geographic poles. At the moment, the north magnetic pole is located at about 78.5°N, 60°W (northeast corner
of Greenland).

FIG. 1-3.   Example of the aurora borealis. (Courtesy of V. Hessler, University of Alaska.)

field, not only ionize the air, but they also produce the beautiful displays of aurora borealis ("northern lights") and aurora australis ("southern lights"). (See Figure 1-3.) Aurora are observed most frequently between 20° and 30° of latitude from the *geomagnetic* poles.

The emanations of light that produce the aurora come from atoms and molecules that have been energized as the result of the collision of incoming solar particles with the atmospheric gases. In fact, much of what is known about the constituents of the outer reaches of the atmosphere has been deduced by examination of the auroral spectrum of light. Even in the absence of aurora, excited atoms and molecules at altitudes above 80 km or so actually emit light continuously, although most of the time the light is very feeble (a brightness of 1/1000 or less of that from aurora). This faint luminescence of the sky, which exceeds the contribution made by stars, is known as *airglow*.

Below the ionosphere, the relative number of electrically charged particles in the atmosphere is so small that the magnetic field of the earth has very little effect on atmospheric behavior. Within the ionosphere, however, the magnetic field has a considerable influence on the motions of the large numbers of electrons and ions. Above a certain altitude, the magnetic field is almost entirely responsible for determining the motion of charged particles (such as those that produce auroras). This outermost region of the atmosphere is therefore called the *magnetosphere*. The outer boundary of the magnetosphere is about 57,000 km above the earth's surface on the side facing the sun but fans out in a broad tail hundreds of thousands of kilometers on the side away from the sun (Figure 1-4). This elongated shape of the magnetosphere is produced by the solar "wind" (or solar "plasma")—a stream of mostly protons and electrons emitted by the sun—which acts to flatten the earth's magnetosphere on the side facing the sun. In 1958, it was discovered that some energetic protons and electrons are trapped within the magnetosphere. The doughnut-shaped regions of high concentration that surround the earth are now known as Van Allen radiation belts. The inner belt is centered at about 3200 km above the earth, while the outer one is located at about 16,000 km (Figure 1-5). It was

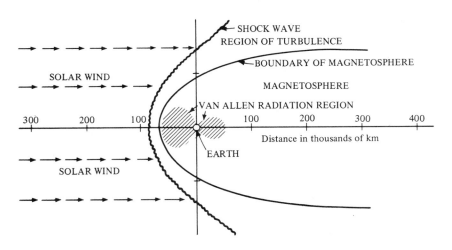

FIG. 1-4.   The magnetosphere.

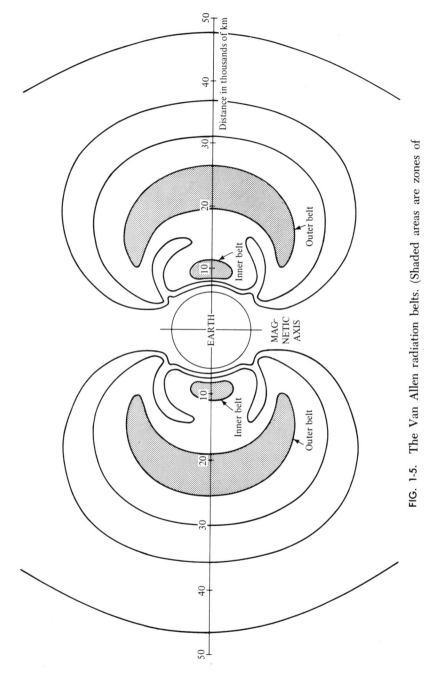

FIG. 1-5. The Van Allen radiation belts. (Shaded areas are zones of maximum intensity.)

15

once feared that the radiation in these regions would be extremely hazardous for astronauts, but it is now believed that transit through them on the way to outer space would be so brief that the astronauts would be protected from possible danger.

1-3. Temperature Distribution in the Vertical

The mean vertical temperature distribution shown on the left in Figure 1-1 is still another basis for dividing the atmosphere into shells or layers. In the lowest of these layers, the *troposphere*, the temperature decreases with height, on the average, at the rate of 6.5°C/km (3½°F/ 1000 ft). In this layer, vertical convection currents, induced primarily by the uneven heating of the layer by the earth's surface, keep the air fairly well stirred. Practically all clouds and weather, and most of the dust and water vapor of the atmosphere are found in this turbulent layer. Its upper boundary, called the *tropopause*, is at an average elevation of about 10 km, but varies with time of year and latitude, and even from day to day at the same place. Typically, the tropopause is at an elevation of 15 or 16 km over the equator, about 10 km over temperate latitudes, and only 5 or 6 km over the polar regions. It tends to be higher in summer than in winter.

In the *stratosphere*, whose upper boundary lies at about 50 km, the temperature is first constant, then increases with height, reaching a temperature at the *stratopause* which is not much cooler than that at sea level. The clouds and the vertical convection currents of air formed near the earth's surface do not usually penetrate very far into the stratosphere. The air in this layer is dry, there generally being less than 0.02 grams of water vapor for every kilogram of air (compared to up to 40 grams per kilogram that can occur near sea level). The absorption of ultraviolet light by ozone in the stratosphere is largely responsible for the increasing temperature with height. Because the ozone is not distributed uniformly over the globe and varies with season (Table 1-5), the temperature within the stratosphere varies considerably both in space and time, and this, in turn, leads to strong seasonal wind circulations within this layer.

The vertical cross section of Figure 1–6 gives some idea of the vertical and latitudinal distribution of temperature in the troposphere and lower stratosphere during one month of the year.

What little water that does exist in the stratosphere is almost invariably in gaseous form. However, on rare occasions, clouds have been observed well within the stratosphere; they may be composed of liquid

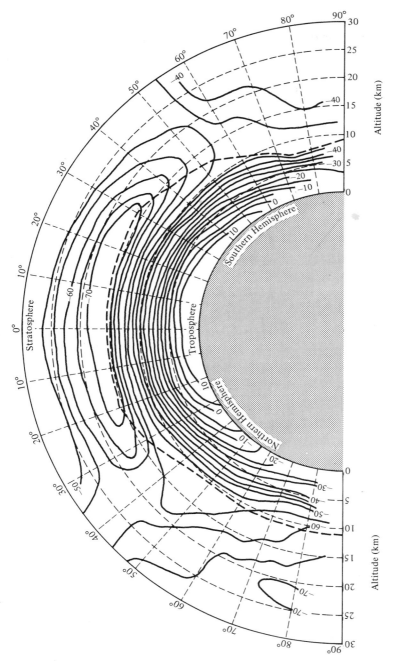

FIG. 1-6. Vertical cross section of average temperature distribution along 75°W longitude for January, 1958.

17

TABLE 1-5

Monthly Variation of Total Ozone (Depth in Centimeters, at Standard Sea Level Temperature and Pressure)[*]

Latitude	Months					
	Jan.	March	May	July	Sept.	Nov.
90°N	0.33	0.46	0.37	0.32	0.27	0.28
70°N	0.34	0.45	0.40	0.34	0.29	0.29
50°N	0.32	0.38	0.37	0.33	0.30	0.29
30°N	0.27	0.29	0.30	0.29	0.27	0.26
10°N	0.23	0.24	0.26	0.25	0.24	0.23
10°S	0.23	0.24	0.24	0.24	0.24	0.24
30°S	0.27	0.26	0.28	0.29	0.32	0.29
50°S	0.31	0.29	0.32	0.39	0.40	0.37
70°S	0.31	0.31	0.34	0.43	0.43	0.38
90°S	0.31	0.30	0.34	0.43	0.41	0.34

[*] The depth of the layer that would result if all of the atmosphere's ozone could be brought to sea level under standard pressure and temperature.

or solid water particles, although no one is really sure. These high-altitude clouds (20–30 km) are best seen when the sun is a few degrees below the horizon and the light above the earth's shadow illuminates them. They often display the entire spectrum of colors, giving rise to the name of *nacreous or mother-of-pearl clouds*. They are seen most often in winter over Scandinavia and Alaska.

The *mesosphere* is the zone between 50 and 85 km in which the temperature again decreases with height, reaching about −95°C at the *mesopause*, which is the coldest point in the atmosphere. The average decrease of temperature with height is about 3°C/km, about half the rate of that of the troposphere. This layer is evidently one of transition between two energy-absorbing layers: (1) the very high atmosphere above 90 km, where dissociation and ionization of O_2 and N_2 are accompanied by heating, and (2) the stratosphere below it, where ozone reaches its peak concentration. As will be demonstrated in a later chapter, vertical mixing of air leads to a fairly sharp decrease of temperature with height, as in the case of the troposphere. The moderate decrease observed in the mesosphere indicates that there is some vertical movement of air in this layer. Observations that are available indicate that the horizontal winds are quite erratic and sometimes reach extremely high speeds.

The mesosphere evidently is extremely "clean" compared to the troposphere. There is little evidence of particulate matter of any kind except for one notable but rare phenomenon: the occurrence of *noctilucent* clouds. These high-level clouds (70–90 km) can only be seen during twilight, although they may occur, unobserved, during the day. In fact, they are best seen when the sun is between 5° and 8° below the horizon; it may be for this reason that these clouds are observed most often at high latitudes, where there are long twilight periods.

No one has yet determined the composition of these clouds. Although they resemble thin cirrus (ice crystal) clouds (see Figure 1-7 and Appendix 2), it seems unlikely that they could be composed of water droplets or ice crystals. One theory is that they are composed of very fine cosmic dust that accumulates near the top of the mesosphere. Just how meteoroid material becomes concentrated in sufficient quantity to form a visible cloud is not yet clearly understood.

Above the mesosphere, the temperature increases rapidly at first, and then more slowly, with height. This hot layer is known as the *thermosphere;* however, do not be misled by the term "hot." One would not feel the heat that is implied by 1200°C at an altitude of 400 km (Figure 1-1) simply because, although the individual particles have energies that are representative of such a temperature, there are so few

FIG. 1-7. Noctilucent clouds. (Courtesy of B. Fogle, Geophysical Institute, University of Alaska.)

gas particles at such altitudes that the total energy transmitted to a body through collisions is indeed quite small compared to the energy of much "colder" molecules at sea level. At heights above 500–600 km, the density of particles is so low that collisions among them are infrequent and some of the particles can escape the gravitational pull of the earth. This zone, which marks a transition from the earth's atmosphere to the very thin interplanetary gas beyond, is called the *exosphere*.

PROBLEMS

Problems marked with an asterisk (°) are the most challenging.

1. Compute the average mass of an air molecule from the data given in Figure 1-1.

2. A light bulb has an air density of about 10^8 molecules/cm³ At what altitude above sea level is such a vacuum achieved naturally?

3. If you were to construct a scale model of the earth and its atmosphere, starting with a 1-m diameter globe, how far from the surface would the following extend?
    (a) Mt. Everest;
    (b) the level at which 99 per cent of the atmosphere is found;
    (c) the level of the tropopause.

4. The average person inhales about 500 cm³ of air 15 times per minute. From Table 1-4, how many grams of dust did a typical city dweller inhale each day? From Table 1-3, estimate the average mass of the particles and then the number of particles inhaled each day.

5. Make a list of at least a dozen ways in which the atmosphere—its constituents and its motions—affect man and his activities.

6. As far as "weather" is concerned, which of the atmosphere's gases are most important? What role(s) does each play in the weather-making processes?

7. Explain in your own words the reasons for the existence of the *heterosphere*.

°8. Imagine a kilogram of air in which 5 g of water vapor are condensed at constant pressure. By how much would the temperature of the air be raised by the latent heat? (The specific heat of air is about 0.24 cal/g/°C. Take the latent heat to be 600 cal/g.) Would you expect such heat derived from condensation to be significant to atmospheric processes?

°9. The following is a list of just a few of the optical and acoustic phenomena produced by the atmosphere. How many of these can you explain?

(a) variations in the audibility of sound; e.g., between day and night;
(b) variations in the distance that objects and lights can be detected with the eye;
(c) the blueness of the sky;
(d) mirages;
(e) halos;
(f) rainbows;
(g) scintillation (twinkling) of stars and of terrestrial light sources and objects;
(h) twilight;
(i) the rumbling of thunder.

2

Atmospheric Measurements

2-1. Characteristics of Gases

A complete description of the physical state of the atmosphere at any moment requires the measurement of dozens of quantities. This is evident from the discussion of the characteristics of the atmosphere in the last chapter. There are significant variations, both in space and time, of the constituents of the atmosphere (gases, dust, and liquid and solid water particles); of electrical charge; of magnetic field; and of heat content. All of these factors play some role in the energy budget of the atmosphere but some, such as the air temperature, the changes in state of water, and the air motion, account for such a large proportion of the total energy and es-

sentially all of everyday "weather," that they deserve special consideration in this chapter.

Since the atmosphere is composed mostly of gases, we shall briefly
review the behavior of gases and the measurements that are used to
describe their physical state. First of all, matter can exist in any of three
states: solid, liquid, or gas. Almost all substances on the earth occur
naturally in only a single state. The gross characteristics of each of the
three states are quite familiar: Solids resist changes in their shape and
do not flow, while liquids and gases are easily deformed and do flow
(hence, they are known as *fluids*); the space occupied by a solid or a
liquid is not easily altered, but a gas spreads out to fill the entire volume
available to it (we say that gases are compressible). These characteristics can be explained largely in terms of how closely the molecules of
the substance are bound: In a solid, the molecules are locked into position and their motion is restricted to oscillations over short distances
from their mean positions; the molecules in a liquid have considerable
freedom of movement, but they are bound to the bulk of the liquid with
sufficient force so that they cannot significantly increase the mean distance between individual molecules; in a gas, the adhesion between
molecules is weak, the molecules are relatively far apart, and they can
move about with comparative freedom throughout the volume to which
the gas is confined.

We shall be concerned primarily with the "gross" or "bulk" properties of matter, particularly those of gases. Although we may occasionally mention molecules and atoms in the way of explanation, we shall
be interested primarily in populations of molecules and atoms and how
they behave as groups.

One such gross property of a gas is the space that a given number of
gas particles may occupy. At sea level, there are about 25×10^{18} molecules of air in each cubic centimeter (about the volume of the tip of
your small finger up to the base of the nail).* The number of molecules
per unit volume is referred to as the molecular *density* of the gas. More
commonly, we refer to the total mass of the molecules, rather than their
number, and the density can then be stated as the number of grams
contained in each cubic centimeter. Near sea level, 1 cm³ of air contains a mass of about 1.2×10^{-3} g (0.0012 g), so that the density is
1.2×10^{-3} g/cm³ (about 0.08 lb, or 1.2 oz per cubic foot).

Another bulk property of fluids is *pressure*, which is defined as the
force per unit area exerted on any surface being bombarded by the

---

* The molecule itself has a volume of only about 6×10^{-24} cm³, so that the
"open" space is about 10,000 times greater than the "occupied" space.

fluid's moving molecules. Held to the earth by gravitational attraction, the earth's atmosphere has a cumulative force or weight per unit area averaging 14.7 lb/in.^{2} at mean sea level. In the cgs system of units, the standard sea level pressure is 1,013,250 dynes/cm^2. Since the dyne/cm^2 is such a small and "wordy" unit, the "bar," which is equal to a million dynes/cm^2, has been introduced as a pressure unit. For meteorological purposes, the *millibar* (abbreviated mb and equal to 1/1000 bar) is most widely used. The pressure of one standard atmosphere is thus 1013.25 mb.

Gases are easily compressed, a fact that is illustrated by the pressure distribution with height given in Figure 1-1 and in the table of Appendix 8. In water, the pressure increases almost exactly in proportion with depth below the water surface, but not so in the atmosphere. As can be seen from the table, one must ascend 2500 m for the pressure to fall off 25 per cent (about 250 mb) from its sea level value, but over 3000 m more for another 25-per cent drop, and another 4800 m for an additional 25 per cent; in the layer between sea level and 5000 m, the pressure decrease is 470 mb, but in the layer between 30,000 m and 35,000 m, the decrease is less than 3 mb. In other words, the pressure falls off with increased height much more rapidly at low altitudes than at high altitudes. Evidently, the air near the bottom of the atmosphere is compressed by the weight of the air resting above. This compressibility characteristic of gases is of great significance in atmospheric processes. As will be pointed out in a later chapter, rapid compressions and expansions of air occur naturally in the atmosphere due to vertical displacements, and these are largely responsible for much of the weather.

Gas Laws

We remind the reader of two fundamental laws about the behavior of gases. These relate the properties of temperature, density (or volume), and pressure. (1) *Boyle's law* states that if the temperature of a gas does not change, its density varies directly as the pressure varies. In symbolic form, $p \propto d$, or $p = kd$, where p is the value of the pressure, d is the density, and k is a constant. In other words, at the same temperature, air at high pressure is more dense than air at low pressure. (2) *Charles' law* states that if the pressure within a gas is kept unchanged, its volume will change in proportion to any temperature change that may occur. This means that, for a fixed amount of mass, the density (mass per volume) is inversely proportional to the temperature when the pressure is constant; i.e., the density decreases when the temperature increases, and it increases when the temperature decreases. In other words, at the same pressure, cold air is more dense than warm air.

These two laws can be expressed more conveniently by a single equation:

$$p = RdT$$

R is a constant that is fixed for any particular gas or mixture of gases. (For air having the relative proportions of gases shown on the left side of Table 1–1, R is equal to 0.287×10^7 erg per °K per gram mass.) Note that both Boyle's and Charles' laws are contained in this single expression: If T, the temperature,* is maintained constant, then $p =$ constant $\times d$; or, if p, the pressure, is kept constant, $d =$ constant$/T$.

2-2.   Observations of the Atmosphere

Weather observations of a sort have been made by man since earliest times, but systematic measurements of the elements did not begin until the invention of instruments during the seventeenth and eighteenth centuries. Until the twentieth century, measurements were confined to the air close to the ground. Systematic measurements of most of the earth's atmosphere are scanty even today.

Measurement of the state of the atmosphere is quite difficult. In addition to the usual requirement that an instrument measure accurately whatever it is designed to measure, the meteorological instrument must be rugged enough to withstand the weather elements—the force of buffeting winds, the corrosive action of high humidity and flying dust, the extremes of heat and cold. Another difficulty is the inaccessibility of much of the atmosphere, so that instruments must be built to transmit their measurements to distant ground points, they must be rugged and light enough to be carried aloft by balloons and rockets, and cheap enough to be used in the large quantities necessary to observe the atmosphere. Finally, the meteorological measurements must be "representative"—a difficult objective to achieve considering the enormous size of the atmosphere and the comparatively few observations that can be made. For example, rainfall is measured as the depth of the water in a rain gauge 8 inches in diameter, a measurement which is assumed to represent the average rainfall over an area of many square miles. This is somewhat like assuming that the height of a

---

* In degrees Kelvin. See Appendix 3 for formulas and tables to convert units.

single student taken at random is equal to the average of the entire school.

Temperature

Temperature is a measure of the degree of hotness or coldness of a body; i.e., of the quantity of internal or heat energy that it contains. A thermometer is a device that measures the degree of hotness or coldness of a body on a numerical scale. This is usually done by correlating physical changes in the thermometer with temperature changes. Thus, for example, the increase in volume of mercury with increased temperature is used in the common mercury-in-glass thermometer. Another way to measure the temperature of a body is to relate changes in the properties of the body itself with changes in the temperature. The color of steel in a furnace is a good index of its temperature; the speed at which sound waves travel through air depends on the air temperature.

Almost every type of temperature-measuring device has been used in meteorology, but the expansion type is the most commonly used for observation near the surface of the earth because of its ruggedness and cheapness. The ordinary liquid-in-glass thermometer is widely used in meteorology. With slight modifications, the liquid-in-glass thermometers can be made to register the maximum or minimum temperature during a period of time. The maximum thermometer has a constriction of the bore of the glass tube just above the bulb; as the temperature rises, the mercury is forced through the constriction; but when the temperature falls, the weight of the mercury in the column is insufficient to reunite it with that in the bulb, so that the top of the mercury column indicates the highest point reached. The maximum thermometer can be reset by shaking the thermometer, thereby forcing the mercury through the constriction.

The minimum thermometer contains alcohol in the bore, with a small, dumbbell-shaped glass index placed inside the column of alcohol. The index is kept just below the meniscus of the alcohol column by surface tension. With the thermometer mounted horizontally, when the alcohol contracts, the meniscus drags the index with it; but when the alcohol expands, the meniscus advances, leaving the index at its lowest point. To reset, the index can be returned to the meniscus by merely tilting the thermometer.

Expansion-type thermometers are also used for recording the temperature. Either a bimetal or a Bourdon thermometer is used to move a pen arm that traces its position on a paper chart driven by a clock. The bimetal thermometer is the type ordinarily used in thermostatic control devices. Two strips of metal, having different rates of expansion during a temperature change, are welded and rolled together. The difference in expansion of the two strips causes changes in the curvature of the element as the temperature changes; with one end fixed in position, the other end is free to move the pen arm and indicate the temperature. The Bourdon thermometer consists of a flat, curved metal tube containing a liquid; as the volume of the liquid changes with temperature, the curvature of the tube changes and this can be used to move the pen arm in a fashion similar to that of the bimetallic strip.

The principal use of electrical thermometers in meteorology is in the radiosonde (Figure 2-1) which is attached to a balloon and transmits temperature, pressure, and humidity data by radio as it ascends through the atmosphere. There are two general types of electrical thermometers: (1) the thermoelectric thermometer, which operates on the principle that temperature differences among the junctions of two or more different metal wires in a circuit will induce a flow of electricity, and (2) the resistance thermometer, which is based on the principle that the resistance to the flow of electricity in a substance depends on its temperature. It is the latter that is used in the radiosonde. The ceramic elements commonly used are called thermistors.

Obtaining meaningful air temperatures is not a simple procedure. Air is a poor conductor of heat and quite transparent to radiation, especially in the short wavelengths emitted by the sun. That this is so is quite evident when one moves a few feet from the shade into the sun; even though the air temperature is almost identical, one feels much warmer in the sun. Or, standing near a fire, the side of the person facing the fire "roasts" while the other side "freezes."

Most thermometers are much better radiation absorbers than air. They absorb energy from the sun and other warm objects that passes right through the air. If the thermometer is to measure the *air* temperature, such radiation must be prevented from reaching the thermometer. This is accomplished by shielding the thermometer, at the same time keeping it in contact with the air. This can be done by enclosing the thermometer within a highly polished tube, allowing plenty of room for air to circulate past the thermometer. However, when several temperature-measuring devices are used, it is convenient to house them in a special "instrument shelter" which permits air to pass through. The shelter also serves to keep the instruments dry during rain, since a wet thermometer will generally read lower than a dry one. (See the section on humidity.)

The thermometer should be ventilated artificially when there is little wind, because the conductivity of air is poor ("dead" or stagnant air

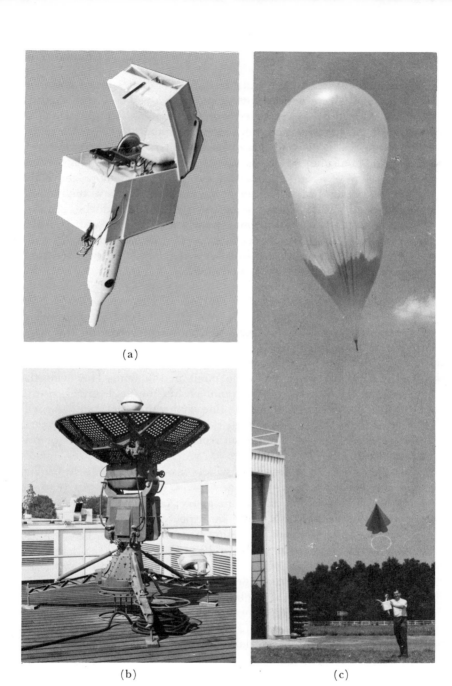

(a)

(b)   (c)

FIG. 2-1. Atmospheric soundings, using the radiosonde. (a) The radiosonde instrument; (b) The ground-radio tracking equipment;
(c) Release of the radiosonde.

is often used for insulation) and the thin layer that encases the thermometer might have a different temperature than the "free" air. By
stirring the air, this layer is mixed with the surrounding air.

Even if the precautions in measuring temperature given above are
taken, there is still the question of how to interpret temperature measurements. On a sunny, windless day, the temperature of the air within
an inch of two of a cement sidewalk can be 30°F or more above the
temperature at the 4-ft level. Even at the same height above the ground,
the temperature differs greatly between the city and the country,
within forests and over open land, along sloping land and flat land.
Differences in the thermometer environment are so important that the
measurements at a single point can rarely be considered representative
of the average conditions closer than 2°F.

Pressure

Pressure* is defined as the force per unit area exerted on any plane
surface. In the case of static fluids, the orientation of the surface will
not affect the pressure. In other words, the force per surface area is
equal in all directions at any one point in the fluid. However, the
pressure can differ from point-to-point within the fluid, giving rise to
a net force per unit area along the line connecting the points. As will
be explained in Chapter 4, which deals with causes of atmospheric
motions, it is the pressure variation from point-to-point that is of primary interest.

In the case of the atmosphere, which has no outer walls to confine
its volume, the pressure exerted at any level is due almost entirely to the
weight of the air pressing down from above; i.e., the force results from
gravitational attraction. As would be expected, the pressure changes
most rapidly in the vertical. In the lowest few kilometers, the pressure
decrease amounts to about 1 mb per 10 m (1 in. Hg/1000 ft). Because
of the compressibility of air, the rate at which the pressure decreases
with height becomes slower at greater heights.

Variations of pressure in the horizontal are much smaller than they
are in the vertical. Near sea level, the change of pressure with distance
rarely exceeds 3 mb per 100 km (3×10^{-4} mb/10 m, or 5 mb/100 miles)
and is usually much less than half this rate. The horizontal variations
in pressure, although small, are able to produce the winds we observe,
as will be shown in Chapter 4. Since pressure measures the weight per
unit area of the atmosphere, variations in pressure along any horizontal

---

* The units of pressure are given in the section on characteristics of gases.

surface (such as sea level) must arise through variations in the average density of the atmosphere; i.e., there must be more molecules in a column of air above a point having high pressure than in one above a point where low pressure is observed.

Pressure also changes with time at a single place. Some of these changes are of an irregular nature, caused by occasional invasions of air having a different mean density. But there is also a quite regular diurnal oscillation of the pressure that causes, on the average, two peaks (at about 10 A.M. and 10 P.M.) and two minima (at about 4 A.M. and 4 P.M.). The difference between maxima and minima is greatest near the equator (about 2.5 mb), decreasing to practically zero above 60° latitude. These regular fluctuations in the pressure are analogous to the tidal motions in the ocean, but in the case of the ocean it is the gravitational pull of the moon and sun that causes the water surface to bulge slightly outward from the earth, while in the atmosphere the daily heating and cooling cycle appears to be by far the dominant cause of the pressure variations. Diurnal wind oscillations accompany the migration of these maxima and minima of pressure around the earth each day; these are hardly detectable at low elevations in the atmosphere, but become quite strong between 80 and 100 km.

The mercurial barometer, invented by Galileo's student, Torricelli, in 1643, is still the fundamental instrument for measuring atmospheric pressure. It is constructed by filling a 33-inch long tube with mercury. The open end of the tube is then inverted and immersed in an open dish of mercury. The mercury in the tube will flow into the dish until the column of mercury is about 30 inches high (at a sea level site), leaving a vacuum at the top (Figure 2-2).

In principle, the barometer is merely a weighing balance, the pressure exerted by the atmosphere on the exposed surface of the mercury in the dish equaling that exerted by the mercury in the tube. Changes in atmospheric pressure are detected from changes in the height of the column of mercury.* It is common to use the height of the column as a pressure unit (millimeters or inches of mercury); conversion to such units as dynes/cm², mb, or lb/in.², can be made as follows: The density of Hg at 0°C is 13.6 g/cm³. (Note that the height of the column of mercury will depend on temperature as well as on pressure, since mercury expands with increased temperature. To obtain the true pressure, one must correct for this mercury expansion.) The mass of a

---

* Mercury is used, of course, because of its high density, which is 13.6 times that of water. Thus, the atmosphere can support a column of water 13.6 times higher than one of mercury. It follows that the maximum altitude to which a suction pump or a siphon can raise water at sea level is about 10 meters.

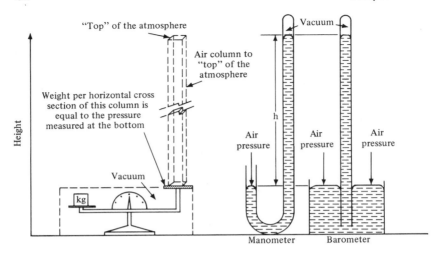

FIG. 2-2.   Principle of the barometer.

column of mercury = mercury density × volume = density × height × cross-sectional area of the tube; its weight, therefore, would be obtained by multiplying by the acceleration of gravity (weight = mass × gravity) and the weight per unit area (pressure) exerted by a column obtained by dividing by the area. Thus, pressure = gravity × density × height. For example, if the height of the column of mercury were 76 cm, the pressure would be $980.6 \times 13.6 \times 76 = 1.0136 \times 10^6$ dynes/cm² or 1013.6 mb. Here, the value of 980.6 cm/sec/sec has been used for gravity; in practice, the gravity value of the particular place should be used.

The aneroid barometer, although not usually as accurate as the mercurial barometer, is more widely used because it is smaller, more portable, usually cheaper to manufacture, and simpler to adapt to recording mechanisms. Its principle of operation is that of the spring balance (Figure 2-3). A thin metal chamber, with most of its air eva-cuated, is prevented from collapsing under the force of atmospheric pressure by a spring. The force exerted by a spring depends on the distance it is stretched. The balance between the spring force and the atmospheric force will thus depend on the width of the chamber. Changes in this width can be discerned by movement of an arm at-tached to one end of the chamber; these deflections are usually magni-fied by levers. If a pen is attached to the arm, the instrument becomes a barograph.

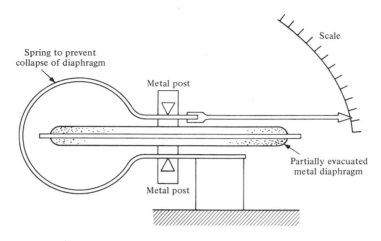

FIG. 2-3.   Principle of the aneroid barometer.

The altimeters used in aircraft and by mountain climbers, surveyors, and others are usually nothing more than aneroid barometers made to indicate altitude rather than pressure. They are designed to give the altitude for the standard ("normal") pressure distribution with height (Appendix 8), and so will often give slightly erroneous readings. For more accurate determinations of altitude, the true density of the air for each altitude increment must be measured and corrections to the indicated altitude computed.

Humidity

The concentration of gaseous water in the atmosphere varies from practically zero to as much as 4 per cent (4 g of water in every 100 g of air). The extreme variability in the amount of water vapor, in both space and time, is due to water's unique ability to exist in all three states—gas, liquid, and solid—at the temperatures normally found on earth. Water vapor is continuously being extracted from the atmosphere through condensation (vapor to liquid) and sublimation (vapor to ice); some of this may fall to the earth as precipitation. Water is continuously being added to the atmosphere through evaporation (liquid to vapor) from the oceans, lakes, rivers, soil, plants, and raindrops, and sublimation (ice to vapor) from snowflakes, glaciers, etc.

The exact amount of water vapor that exists at any place and time is important to the meteorologist because of the significant role water plays in weather processes. First of all, condensation is an important aspect of "weather." Second, water vapor is the most important radiation absorber in the air and thus affects the energy balance of the atmosphere (Chapter 3). Third, the release of the latent heat of condensation is an important source of energy for the maintenance of atmospheric processes.

For a substance such as water to change its phase from solid to liquid or liquid to gas, the forces that bind the molecules together must be broken down. Work must be done in overcoming these intermolecular forces, so the molecules must expend part of the internal energy. The molecules acquire this energy from their environment. It is for this reason that skin is cooled by evaporation of perspiration, and water in a porous water bag is cooled by evaporation through the walls. The energy required to effect a "phase" or state change such as occurs in evaporation is called *latent heat* because it reappears when the change of state is reversed. Thus, to evaporate 1 g of liquid water, approximately 600 cal are required; if the same gram of gaseous water is returned to liquid state, the 600 cal will be released to the environment. A similar thing happens during the ice-to-liquid transition, but the *latent heat of fusion* is only about 80 cal/g.[*] Changes directly between ice and vapor involve a latent heat of sublimation which is the sum of the latent heats of fusion and vaporization; i.e., approximately 680 cal/g.

We refer to the gaseous state of water as "vapor" because it is so easily condensed, but it acts much like any other gas in the atmosphere. The molecules of water vapor move about, occupy space, and exert pressure as do the other gases, except that the amounts of most of the other gases are relatively fixed. The quantity of water vapor in the air can be expressed in a variety of ways. One is the density of water vapor, usually referred to as the *absolute humidity* and expressed as the number of grams of water vapor in a given volume. Normally, there are not more than about 12 g/m³, although as much as 40 g/m³ can occur.

The *partial pressure* of water vapor; i.e., the contribution made by water to the total atmospheric pressure, is another measure that can be used. It is usually expressed in millibars or inches of mercury. Typically, the water vapor pressure does not exceed 15 mb (0.44 in. Hg), although it can reach double or more this value.

The amount of water vapor that can be added to a volume at any given atmospheric pressure and temperature is limited. When a volume

---

[*] The latent heats depend somewhat on the temperature at which the phase changes occur.

has reached its capacity for water vapor, it is said to be saturated, and the volume will accept no more gaseous water. The *saturation vapor pressure,* as this maximum vapor pressure is called, is a function of temperature. (See Figure 2-4.) The first two columns of Table A of Appendix 7 also illustrate how the saturation vapor pressure varies with temperature near sea level. This dependence of the saturation vapor pressure on the temperature is why cooling is so important in producing condensation. For example, if a sample of air having a temperature of 70°F and a water vapor pressure of 0.600 in. Hg were cooled to 45°F, the sample would become saturated at a temperature of about 64°F (interpolating between 60° and 65°F in the table) and further cooling would result in condensation of the excess moisture; when the 45°F temperature was reached, the saturation vapor pressure would be only 0.300 in. Hg, and so half of the vapor would have liquefied. The temperature to which a sample of air must be cooled (at constant atmospheric pressure) to make it "saturated" is called the

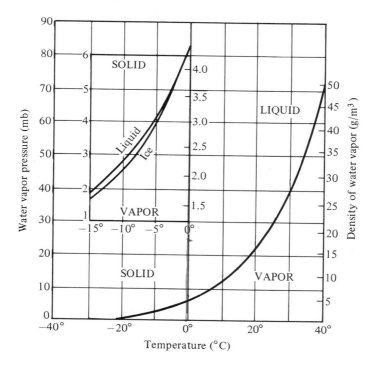

FIG. 2-4. Saturation vapor pressure and density as a function of temperature. (Inset: variation below 0°C.)

dew point. In the example given, the dew point of the sample before condensation began was 64°F; *after* condensation begins, the temperature and the dew point are equal. Thus, the dew point is a direct measure of the water vapor pressure, and the difference between the temperature and the dew point* is a measure of the degree of saturation of the air.

The *relative humidity* is the ratio of the actual vapor pressure to saturation vapor pressure; i.e., relative humidity = actual vapor pressure/saturation vapor pressure. The ratio is usually multiplied by 100 and expressed in per cent. Relative humidity measures how close the air is to saturation (100 per cent indicates complete saturation). In the preceding example, the relative humidity *before* cooling was $(0.600/0.739) \times 100 = 82$ per cent; between 64°F and 45°F, the relative humidity remained constant at 100 per cent. Thus, the relative humidity is very sensitive to temperature change. There is normally a large diurnal change of relative humidity, even when the quantity of moisture in the air is constant, merely because the daily temperature variation continuously changes the saturation vapor pressure.

The most accurate way to measure humidity is to extract all of the water vapor from an air sample (perhaps by passing it through a chemical drying agent) and then weigh the water collected. However, for meteorological observations, this procedure is not practical, since the sampling time is too long and the analytical tools are too complicated. Study of the atmosphere requires almost instantaneous sampling under field conditions.

None of the many techniques used to measure atmospheric humidity is completely satisfactory; here we will mention just a few of the most commonly used instruments. The hair hygrometer is probably the oldest and most widely used of moisture-measuring instruments. Many organic materials such as wood, skin, and hair absorb moisture when the humidity is high, and so they expand. Hair on the human head increases its length by about 2½ per cent as the relative humidity increases from 0 to 100 per cent. The hair hygrometer merely consists of one or more hairs, the changes in length of which are made to move a pointer or, in the case of a hygrograph, a pen.

The psychrometer consists of a pair of ordinary liquid-in-glass thermometers, one of which has a piece of tight-fitting muslin cloth wrapped around its bulb. The cloth-covered bulb, called the wet bulb,

---

* When the saturation temperature is below freezing, the temperature to which the air must be cooled to produce sublimation is referred to as the *frost* point. For example, if the water vapor pressure were 0.135 in. Hg, the frost point would be 25°F (Table A, Appendix 7).

is wetted with pure water and both thermometers are then ventilated. The dry bulb will indicate the air temperature, while the wet bulb will be cooled below the dry-bulb temperature by evaporation. The amount of evaporation, and therefore of cooling, will depend on the degree of saturation of the air. If the surroundings are saturated, there will be no evaporation, and the wet and dry bulbs will read the same. The difference between the dry and wet bulbs, called the *depression of the wet bulb,* is a measure of the degree of saturation of the air. The tables of Appendix 7 can be used to obtain the relative humidity or dew point from psychrometric readings. As an example, if the dry bulb temperature were 60.0°F and the wet bulb 55.0°F, the relative humidity would be 73 per cent and the dew point 51°F. Note that both thermometers must be read to the nearest 0.1°F in order to determine the relative humidity within ±1 per cent or the dew point within ±0.2°F.

An electrical hygrometer is used in the radiosonde. It consists of an electrical conductor that is coated with lithium chloride, which is hygroscopic. The amount of moisture absorbed by the conductor depends on the relative humidity of the air, and the electrical resistance of the conductor is a function of its dampness. Another less common technique for measuring humidity is that of the infrared hygrometer. This relates the depletion of an infrared light beam to the amount of water contained in an air sample.

Clouds

In the discussion on humidity, it was implied that as soon as the concentration of vapor begins to exceed the saturation value, which is largely dependent on temperature, the excess moisture becomes liquid or solid. It is not quite that simple. For condensation or sublimation to occur, there must be a suitable surface available. Dew or frost forms easily on grass, soil, windows, etc., whenever the air temperature reaches the dew point or the frost point. But in the free atmosphere there are no such extensive surfaces.

In pure air (i.e., all gas with no "foreign" particles, water droplets, or ice crystals), condensation or sublimation is extremely difficult to achieve, even under highly supersaturated conditions (relative humidity much greater than 100 per cent). Wilson demonstrated with his famous cloud chamber (the one that became such an important instrument for nuclear physicists) that in completely clean air the relative humidity must approach 800 per cent before water drops will form. For a single droplet to form in clean air, a large number of water molecules* must not only collide with each other but also stick together. The probability of such multiple collisions is extremely small; and even when some molecules do stick to form a small "embryo," thermal agitation of the molecules tends to cause some of the outer molecules to escape. Only after a droplet has reached a critical size (radius greater than about 10^{-5} cm) are the binding forces sufficient to hold more of the molecules that strike its outer surface than the number that escape it. Since molecular speed depends on temperature, this critical size is a function of temperature. Once this critical size is reached, further growth is very possible; but the probability that the hundred million molecules required to produce such a size will not only collide but stick together is extremely small in pure air that is not greatly supersaturated.

Fortunately, in the natural atmosphere there are numerous particles much larger than individual molecules that provide surfaces or *nuclei* to which water molecules can adhere. It is around these that water droplets and ice crystals grow. (Cloud and raindrops are never "pure," despite the myth that they are, although the proportion of "foreign" particles to water is usually very small.) Certain types of particles, such as salts injected into the atmosphere from the sea, attract water molecules to their surfaces, and are said to be *hygroscopic nuclei.* On these, condensation may actually begin well before the air becomes saturated (relative humidity as low as 70 per cent). However, salt nuclei represent only a small part of the total number of particles suspended in the air, especially over land areas, and there are a large number of other types of particles, such as combustion products, meteoritic dust, volcanic material, and soil, that also serve as nuclei. These small particles, most of which have diameters of less than 1 micron (a thousandth of a millimeter) and weigh only about 10^{-15} gram, in concentrations of 10,000 or more per cubic centimeter, are so small that they remain suspended in the air for days at a time. (See Table 1-3.)

There is another interesting property of water: Although we are accustomed to thinking of 0°C as the freezing point, it would be more accurate to call this temperature the melting point. In fairly large volumes of water, such as ponds or lakes, freezing *does* occur very close to 0°C. But in very small droplets, it is possible to achieve a temperature as low as −40°C before freezing takes place. Indeed, in the atmosphere it is common for liquid water drops to exist in clouds at temperatures as low as −20°C.

---

* About 1000 molecules are needed to form a droplet having a diameter of only 4×10^{-7} cm (the wavelength of X rays). Thus, a typical cloud droplet, which has a diameter of about 2×10^{-3} cm, would contain about 10^{14} molecules.

For pure water to become ice, the molecules must arrange them-selves into a particular configuration. Such an arrangement arises through chance. The probability that the correct arrangement will arise somewhere without being pulled apart by the molecular oscillations will, of course, increase with the volume of water involved. The lower the temperature, the smaller is the typical molecular velocity, and thus the greater are the frequency and size of ice-like molecular aggregates in a given water sample. If an ice "embryo" reaches a size sufficiently large so that it can hold its family of molecules, it can then grow, act-ing as a seed or nucleus to which other molecules can adhere. Freezing of the entire sample of water will then proceed very rapidly.

Certain foreign particles in water can also act as nuclei that induce freezing. These tiny particles are called *ice nuclei* or *freezing nuclei*. Like condensation nuclei, freezing nuclei range in size from 10^{-4} to 10^{-3} mm. Not all dust particles serve as ice nuclei, and those that do vary considerably in their effectiveness (the temperature at which they cause freezing). The property that determines effectiveness of a parti-cle or a nucleus is still not completely settled, although it may have something to do with the crystal form. Certain artificial nuclei such as silver iodide particles (which have an atomic arrangement similar to that of ice) have been found to be very effective in causing supercooled drops to freeze.

Fog and clouds are composed of liquid water and/or ice particles suspended in the air. There can be as many as 500–600 particles in each cubic centimeter, although normally there are fewer than half this number.* The particles in nonprecipitating clouds are normally quite small, however (averaging about 0.01 mm in radius and rarely exceed-ing 0.1 mm), and so they fall to earth very slowly. In calm air, a drop-let having a radius of 0.05 mm falls at the rate of less than 1/2 m/sec (1 mph). This maximum fall velocity, known as the *terminal velocity* (Table 1-3), is imposed by air resistance. Most clouds are formed when air is rising (as will be seen in Chapter 4), so that in practice, even in extremely weak upward air currents, drops of this size can be sus-pended in the atmosphere for many hours. In fact, clouds begin to pre-cipitate only when some of the drops within them reach sufficient size to fall through the air with an appreciable velocity. Even when a rain-drop falls earthward, it may not reach the ground. If the air beneath the

---

* Clouds that form over oceans generally have only 30 or 40 drops per cubic centimeter compared to several hundred in "continental" clouds, and drop sizes are typically several times larger in oceanic clouds. The reason for these differences appears to be that in clouds over land there is a greater concentration of nuclei, all of which compete for the available supply of moisture.

cloud is fairly dry and the drop small (diameter less than 0.4 mm), it will be evaporated before it has traveled more than one or two hundred meters. *Virga,* the gray streaks that sometimes trail from beneath clouds, is rain that has evaporated before reaching the ground.

Cloud types. Aside from observation of the internal makeup of clouds, the outward appearance of clouds is of significance to the meteorologist in interpreting the physical processes in the atmosphere, and is often a harbinger of the weather to come. The basic clouds are identified on the basis of their form and approximate height above the ground where they normally occur. The names of the basic clouds are composed of the following roots: *cirrus* (meaning feathery or fibrous); *stratus* (stratified or in layers); *cumulus* (heaped up); *alto* (middle); and *nimbus* (rain). The ten basic clouds are:

High (base above 6 km, or 20,000 ft, generally composed entirely of ice crystals): *cirrus* (Ci), *cirrostratus* (Cs), *cirrocumulus* (Cc).

Middle (2–6 km, or 6500–20,000 ft): *altocumulus* (Ac), *altostratus* (As).

Low (below 2 km or 6500 ft): *stratus* (St), *stratocumulus* (Sc), *nimbostratus* (Ns).

Clouds of vertical development (base usually below 2 km, or 6500 ft, but top can extend to great heights): *cumulus* (Cu), *cumulonimbus* (Cb).

Illustrations of these clouds along with more detailed descriptions of their appearance and formation processes are contained in Appendix 2. Some common adjectives applied to the basic names to further describe particular clouds are:

Uncinus: hook-shaped; applied to cirrus, often shaped like a comma.

Castellanus: turreted; applied most often to cirrocumulus and altocumulus.

Lenticularis: lens-shaped; applied mostly to cirrostratus, altocumulus, and stratocumulus; occurs when air currents are undulating sharply in the vertical, as is sometimes the case on the lee side of mountains.

Fractus: broken; applied only to stratus and cumulus.

Humilis: lowly; poorly developed in the vertical; applied to cumulus.

Congestus: crowded together in heaps, like a cauliflower; applied to cumulus.

As will be shown in Chapter 4, almost all clouds result from the rapid cooling of air when it ascends. In stratiform clouds, the motion in the vertical that produces them is generally small (less than 20 cm/sec), while in cumuliform clouds, the upward and downward velocities are much stronger (up to 30 m/sec or more). (See Appendix 2.)

Fog is merely a cloud in which the observer is immersed. It is thus a suspension of small water droplets. When the fog is formed of tiny ice crystals, it is known as *ice fog*. *Mist* (or haze) often precedes and follows fog. Mist is formed of very small droplets or wet hygroscopic particles. The air does not feel "wet" as it does in fog.

Although most clouds and fogs are produced by the cooling of air, there are some instances in which the rapid addition of moisture to the air is also very important. A common example of this is, of course, the moist exhaled breath on a very cold day. A similar phenomenon is that of *steam fog*, which forms when air moves across a much warmer water surface. The clouds that form over geysers is another example. Aircraft flying at high altitudes within the troposphere where the air temperature is low often leave trails of clouds behind them. These condensation trails (called "contrails") are formed from the water vapor created during the combustion of the fuel. (The addition of condensation nuclei from the engines may also be a factor.) With the greatly increased number of high-flying jets in recent years, there has been some speculation that the widespread creation of these clouds may have an effect on the weather by intercepting solar radiation and/or supplying condensation nuclei at high elevations in the atmosphere. (See Chapter 10.)

Precipitation

Raindrops are usually between one and several millimeters in diameter. Since the average drop diameter in a nonprecipitating cloud is only about 0.02 mm, this means that many cloud drops have increased their volume by a factor of 1,000,000 by the time they fall out of a cloud. Such growth cannot be explained merely by further condensation of water vapor on existing water particles, because of the inordinately long time that this would require. Since clouds have been observed to form and begin to precipitate in an hour or less, further growth after initial formation of droplets through condensation must result from drops and/or ice crystals combining with each other. Just how this is accomplished is still not completely agreed upon, but the most probable mechanisms are the following:

1. *Collision and coalescence of particles.* The drops formed in a cloud are not all of the same size. Due to differences in the rate

at which condensation proceeds in different parts of a cloud—sometimes separated by very small distances—the largest drops in the cloud may have diameters several times greater than the smallest. As the air swirls about, the larger drops, because of their greater mass, have more inertia than the smaller ones. Therefore, these large drops tend not to follow exactly the same path as the small ones and, as a result, collide and often coalesce (combine) with the small ones. Repeated collision and coalescence by a drop may cause it to grow so large that it splinters into several drops, which in turn grow by collision and coalescence, thus producing a "chain reaction" of raindrop growth.

2. *Growth of ice crystals.* Frequently, especially in the middle latitudes, the uppermost layers of clouds will be composed of ice crystals, while the lower layers contain supercooled (at temperatures below $0°C$) liquid drops. Through stirring within the cloud, or because of different fall velocities of the particles, ice crystals and supercooled drops become mixed. At the same temperature, the saturation vapor pressure over a liquid surface is greater than that over an ice surface (see insert of Figure 2-4), and the ice crystals will therefore grow at the expense of the water drops. This growth mechanism, which is called the Bergeron process after the Swedish meterologist who first suggested it, is believed to be quite important in the initiation of precipitation, although further growth most likely involves the collision-coalescence of particles described in the previous paragraph.

The importance of the coexistence of ice crystals and supercooled water drops in the initiation of precipitation forms the basis for many modern cloud-seeding experiments. In a cloud that contains few or no ice crystals, dry ice introduced into the cloud may cool enough drops to their freezing point to produce crystals for later growth by the Bergeron process. Supercooled droplets can also be induced to freeze through injection of certain types of materials, such as silver iodide which, as mentioned earlier, act as freezing nuclei.

Precipitation types. The only difference between drizzle and rain is the size of the water droplets. The diameter of the former is generally less than 0.5 mm (.02 in.). (See Table 1-3.) The principal solid forms of precipitation are:

Snow: Ice crystals that have grown as they traverse the cloud. (See Figure 2-5 for some of the diverse symmetric forms they develop.)

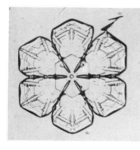

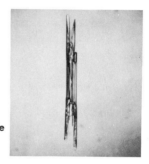

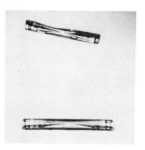

Graupel

Dendritic crystal

Hexagonal plate

Needle

Sector-like crystal

Sheath-like crystal

FIG. 2-5. Some forms of snow crystals. (Courtesy of C. Magano, Hokkaido University, Japan.)

At temperatures greater than about −5°C, crystals usually stick together to form snowflakes.

Freezing rain and drizzle: Rain or drizzle that freezes on impact with the ground or objects.

Snow pellets: White, opaque, spherical or conical grains of ice, having a diameter of 2–5 mm (0.1-0.2 in.). Usually occur in showers when the temperature near the ground is close to 0°C.

Snow grains: Very small (diameter <1 mm), white, opaque, somewhat flattened grains. Occur in nonshowery-type precipitation.

Ice pellets: ("Sleet") Transparent or translucent, quasi-spherical ice, having a diameter <5 mm. These originate either as raindrops or snowflakes that have melted en route to the ground and are frozen as they traverse a cold air layer near the ground.

Hail: Small balls or chunks of ice with a diameter of 5–75 mm (0.2–3 in.) or more that fall from cumulonimbus clouds. These destructive stones are formed by the successive accretion of water drops around a small kernel of ice moving through a thick cloud; as drops are frozen onto the nucleus, they may form new shells, each a millimeter or more thick, so that many hailstones acquire an onion-like cross section (Figure 2-6).

(a)

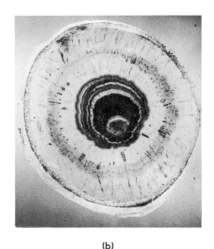

(b)

FIG. 2-6. (a) Hailstones; (b) Cross section of a hailstone. (Courtesy of R. List, SLF.)

Measurement of precipitation. For practical purposes of water supply, meteorologists are concerned with measuring the amount of water reaching the earth's surface. This is done by sampling the depth of water that would cover the surface if the water did not run off or filter into the soil. A rain gauge is merely a collection pail with a ruler to measure the depth of water. The depth is usually measured in increments of a hundredth of an inch or a millimeter. Solid forms of precipitation are melted and the equivalent liquid depth recorded. In the case of snow, which may remain on the ground for a long period of time and thus serve as a natural water reservoir, the snow depth is also of interest. Normally, the ratio of snow depth to liquid equivalent is about 10 to 1, although sometimes it is as much as 30 to 1.

Precipitation amounts are extremely variable from place to place, even during a single storm, so that the problem of adequate sampling over an area is a serious one. In most places of the world, not more than one 8-inch diameter rain gauge is installed in every 100 square miles, which is a ratio of areas of about 50 in.^{2} to $100 \times (5280 \times 12)^2$, or approximately $1:10^{10}$; this is somewhat like taking a single hair from one Californian's head to characterize the hair of everyone in the state. In mountainous areas especially, amounts may vary by a factor of two or three in a distance of less than 10 miles. (Along the northeast slopes of Hawaii, the annual rainfall varies from 15 in. to over 300 in. in a distance of about 15 miles.) Interpretation of measured amounts must be done with considerable care. Even at the same point, annual amounts vary greatly, especially in semi-arid climates; the year-to-year amounts can easily fluctuate by 50 per cent or more of the long-term average annual precipitation. For this reason, claims by rainmakers that they have increased the rainfall in a single year by some precise figure, like 10.5 per cent, should be viewed with a great deal of skepticism.

The Hydrologic Cycle

Man has always been highly dependent on water supply. Today, the increasing growth and concentration of population and of industry are increasing the demands on an essentially fixed world water supply and we have become more concerned than ever with the need to carefully budget our water resources.

One of the hydrologist's tasks is to study the distribution of the world's water supply. Most of the earth's total water is contained in two natural "reservoirs"—the oceans (97 per cent) and the ice fields or glaciers (2 per cent). The combined resources of the continents (lakes, rivers, soil, and underground water) represent only about 0.8 per cent of the total. Even this relatively small proportion of water over the continents would soon drain into the oceans if there were no transport of water from the oceans to the continents. The atmosphere, although containing only about 0.2 per cent of the total water on the average, provides the "hydrologic link" from oceans to land.

At any moment, the atmosphere carries a seemingly small amount of water. For example, even if all of the water vapor could be precipitated, an average depth of only about 1 inch of liquid water would result (Figure 2-7 and Table 2-1).

But new water is constantly being added through evaporation (at the rate of some 1.4×10^{17} gallons per year, 80 per cent from the seas) as well as extracted through precipitation. Within the moving currents of air, the amount of water passing over an area during a period of time can be considerable. For example, even over arid Arizona, the atmosphere carries as much water in a single week in July as flows in the Colorado River during an entire year. Actually, even during heavy precipitation, only a small percentage of the total water in a vertical air column is precipitated; the intense inflow of air into a storm (as in a hurricane) more than compensates for the low rate of moisture extraction.

TABLE 2-1

Average Total Water Content (cm) *in the Atmosphere as a Function of Latitude, Season, and Altitude*

	Summer or wet season				Winter or dry season			
Latitude:	0	30	45	60	0	30	45	60
Above:								
sea level	5.4	4.3	2.6	2.2	3.2	1.6	0.8	0.5
1000 m	3.0	1.9	1.6	1.2	2.0	1.0	0.5	0.3
2000 m	2.0	1.5	1.0	0.8	1.0	0.6	0.3	0.2
3000 m	1.0	0.8	0.5	0.4	0.6	0.3	0.2	0.1

The annual rainfall in the conterminous United States ranges from as little as 4 inches in the deserts of the Southwest to more than 100 inches in the Cascades of Washington and Oregon. The average annual precipitation is about 30 inches. Of this amount, about 70 per cent goes directly into the atmosphere through evaporation from the soil, lakes, and rivers, and through transpiration from vegetation. The remainder runs off into the rivers (about 10 per cent via underground streams and the rest along the surface), eventually emptying into the oceans.

Hydrology deals with all factors that affect water supply over land: precipitation, storage in soil and underground reservoirs, runoff, evaporation, and the return of the moisture to the atmosphere and oceans. In the United States, a network of more than 13,000 precipitation gauges measures rain and snow; more than 300 evaporation pans are used to estimate evaporation. Storage of water in snowfields is obtained by measurements of snow depth. The flow rate and depth of rivers are also measured. From such information, the hydrologist attempts to

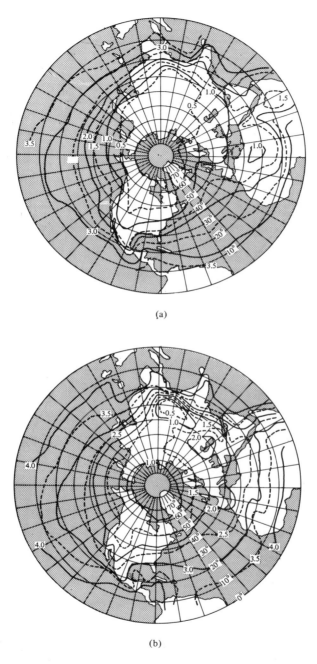

FIG. 2-7. Total precipitable water in the troposphere, g cm^{-2}. (a) winter; (b) summer. (After Adem. MWR.)

47

anticipate destructive floods, drought and future water supply, and to design reservoirs, sewer systems, bridges, etc.

Wind

Air in motion, or *wind*, is the "equalizer" of the earth's atmosphere. By transporting heat, moisture, pollutants, etc., from one place to another, it acts to redistribute the concentration of these quantities. Chap-

FIG. 2-8. Windvane and cup anemometer. (Courtesy of ESSA.)

ter 4 will discuss the causes of winds; at this point, only some of the general characteristics of wind will concern us.

Although air moves up and down as well as horizontally, the speed of vertical displacements is usually a tenth or less of the horizontal component. Even though it is quite small, the vertical component is very important. As we shall see later, it is the up-and-down motion of air that is principally responsible for the formation and dissipation of clouds in the atmosphere. Only the *horizontal* component of the wind is measured on a regular basis, while the much smaller vertical component must be computed from relationships between it and the changes of the horizontal wind in space.

Many instruments are used to measure the horizontal wind velocity near the surface of the earth. The windvane (Figure 2-8) is a very old device for indicating wind direction. Because it points into the wind, it is customary to designate wind direction as that direction from which the air comes. Thus, when the air is moving from northwest (315°) to southeast (135°), the wind direction is said to be northwest (315°).

Many types of *anemometers* exist for measurement of wind speed. The cup anemometer is probably the most widely used. It consists of three or more hemispherical cups clustered around a vertical shaft. Air striking concave sides of the cups exerts more force than that hitting the convex sides, causing the cups and therefore the shaft to turn. The number of rotations per unit time is a measure of the wind speed.

Air flow is retarded by friction with the ground and deflected by obstacles, so that the position of wind-measuring instruments must be carefully considered. At an airport, for example, both the wind speed and direction atop the control tower may be considerably different from the speed and direction at the end of the runway. Typically, the wind speed increases rapidly with height near the surface, so that the height of an anemometer will greatly influence the speed recorded. Unfortunately, there is no uniformity of height for anemometers, although arbitrary standards have been set.

Gustiness and the diurnal wind variation. Anyone who has watched a windvane oscillate, a cup anemometer alternately increase and decrease its rotation speed during brief periods of time, or a flag flutter in the wind can attest to the normal unsteadiness of the wind. An example of such fluctuation can be seen from the recording of the wind direction and speed of Figure 2-9. These velocity changes are attributed to the fact that air normally does not move in stright lines, but in tortuous paths. We call such erratic air flow patterns *turbulent*. Successive particles passing a single point in space may have had distinctly different

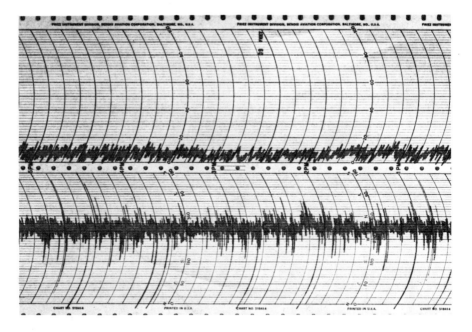

FIG. 2-9.   Example of fluctuation of the wind. (Upper trace is speed,
            lower one is direction.)

histories, some having been most recently at higher elevations than the
point, others at lower elevations. Normally, those coming from higher
up will arrive with relatively high velocities, while those coming from
lower elevations will have relatively low velocities. The result will be a
gusty wind velocity at the observation point.

Turbulent and gusty winds are most pronounced during midday,
when the vertical stratification of the air (Chapter 4) is least, and
weakest during the night, when the stability of the air tends to suppress
vertical displacement of the flow. For the same reason, the winds near
the surface tend to be stronger in the afternoon than they are during
the night.

2-3.   Upper-Air Observations

Sounding Techniques

Winds at levels above the reach of ground-based instruments are
measured by tracking helium- or hydrogen-filled balloons. The horizontal displacements over short intervals of time as the balloon ascends

give the velocity. The changes in position of the balloon may be determined by any of these methods: (1) optically, by the use of a theodolite (similar to a surveyor's transit); (2) by reflection of radio waves (radar) from a target carried by the balloon; (3) by tracking of the radio signal transmitted by a radiosonde carried by the balloon (Figure 2-1).

Systematic measurements of meteorological conditions in the free atmosphere high above the surface began around the turn of the century. Until 1938, when the radiosonde came into use, the sounding instruments had to be retrieved before the data became available. Instruments recording temperature, humidity, and pressure were carried aloft by balloons, kites, and airplanes. In the case of a free balloon, the instrument dropped to the earth on a parachute, and the processing of the data had to wait until some finder returned the instrument. Kites were extremely laborious to handle and they rarely reached heights greater than 3 km. Airplane soundings were expensive, and, at least in the early days, could not provide data to the altitudes desired, or during periods of severe weather.

The radiosonde, carried aloft by balloons, transmits its measurements by radio back to a ground station. The radiosonde used in the United States consists of a lightweight, cheap radio transmitter that emits a continuous signal. The temperature- and humidity-measuring elements control the frequency or amplitude (intensity) of the audio output of the radio signal. An aneroid barometer cell, moving a contact arm across a series of metal strips, alternately connects temperature and humidity into the circuit. By setting consecutive contacts for known pressure intervals, the temperature and humidity are recorded as a function of pressure. The altitude can be computed from the measured vertical distribution of temperature, humidity, and pressure. The radiosonde is now the principal tool of the meteorologist for systematically observing the conditions of the lowest 30 km of the atmosphere. Exploration of higher levels has been accomplished principally by rockets.

Radar

Electromagnetic waves differ from waves of a vibrating string or water waves in that the medium through which electromagnetic waves travel is not needed for their propagation. Nevertheless, any matter that is in their path does affect their motion. The effect of obstacles depends on their size relative to the wavelength. What occurs is very similar to what happens to a water wave when it encounters a floating vessel: If the vessel is very small compared to the wavelength, the wave passes on relatively undisturbed, merely causing the vessel to oscillate up and down. In contrast, if the vessel is very large compared to the wavelength, the wave will be bounced off like a rubber ball; the wave is said to have been *reflected*. In between these two extremes, the obstacle absorbs some of the energy of the incident waves and becomes a source of new waves—ripples that radiate in all directions from the object; this phenomenon is called *scattering*.

Radio waves are both reflected and scattered by obstacles in their path. This fact is the basis for *radar*—"*r*adio *d*etecting *a*nd *r*anging." A transmitter emits a narrow beam of short-wavelength energy in short pulses. Between pulses, a receiver listens for any waves that may be reflected or scattered back to the antenna. From the elapsed time between the transmitted pulse and its return, the distance of the "obstacle" can be easily calculated, its direction being that toward which the antenna is pointed.

Since World War II, radar has come to be used more and more to probe the atmosphere for liquid and solid water particles. The amount

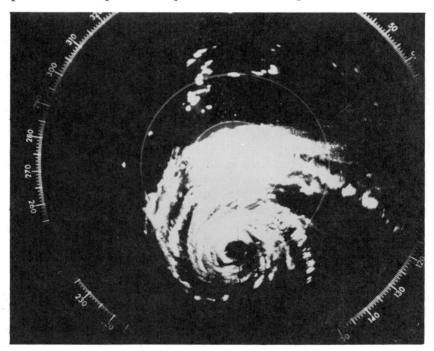

FIG. 2-10. Hurricane Donna (September 9, 1960) as seen by radar. (Courtesy of G. E. Dunn, ESSA.)

of backscatter from these particles depends strongly on their size and the wavelength of the radio waves. For example, with a radio wavelength of close to 1 cm, raindrops and snow crystals having a diameter of 1 mm or more can be easily detected. The much smaller cloud droplets and ice crystals (diameter <0.2 mm) scatter so little energy that only extremely powerful and sensitive radars can "see" them, except at very short wavelengths (<1 mm). But sufficiently powerful radar that generates and receives such extremely short wavelengths is very difficult to build. Radar is now widely used in meteorology for the practical problem of locating precipitation areas and tracking their movement. It is an especially valuable tool for locating and following severe thunderstorms, which are often "lost" in the coarse grid of weather stations. Radar may also be used to obtain an estimate of the areal distribution of rainfall. Figure 2-10 shows a hurricane as it appeared on the radar screen.

The maser and laser° have recently been used as tools for locating extremely small particles. Operating at very short radio waves (maser) or in the visible portion of the spectrum (laser), they have been used successfully to locate layers of haze and dust at high levels in the atmosphere.

Satellites

Observations of most of the earth's atmosphere are entirely inadequate. About 70 per cent of the earth's surface is covered by oceans† and a large proportion of the rest is dominated by mountains, snow, deserts, and jungles. Even in populated areas, the density of weather-observing stations permits the construction of only a very "coarse-grained" picture of the atmosphere.

The launching of weather satellites promises to improve this situation somewhat. Equipped with television cameras (Figure 2-11), they transmit to earth pictures of the clouds as seen from above. These photographs have been valuable, especially over the oceans, to pinpoint the locations of storms. On a few occasions, hurricanes that were undetected by the low-density oceanic network of stations have been uncovered by Tiros weather satellites. Measurements of the infrared radiation from the earth's surface and clouds also promise to yield in-

---

° Microwave or light amplification by stimulated emission of radiation.

† Vessels provide some information over the oceans, mostly along the well-traveled trade routes. Attempts are being made to supplement these data with automatic floating weather stations, such as that shown in Fig. 2-12.

creased knowledge of the earth's energy balance (Chapter 3). A list of United States Meteorological Satellites and some of their characteristics is presented in Appendix 10.

A word of caution should be given regarding extravagant claims that satellites have advanced the science of meteorology tremendously. Photographs tell us very little about the complex physical processes going on within the atmosphere, since clouds are merely visual manifestations of the interplay of these processes. Smoke seen from a great distance may indicate that there is a fire, but it does not tell what is burning, or why. The principal value of satellite photographs taken at altitudes of several hundred miles has been in "finding the smoke."

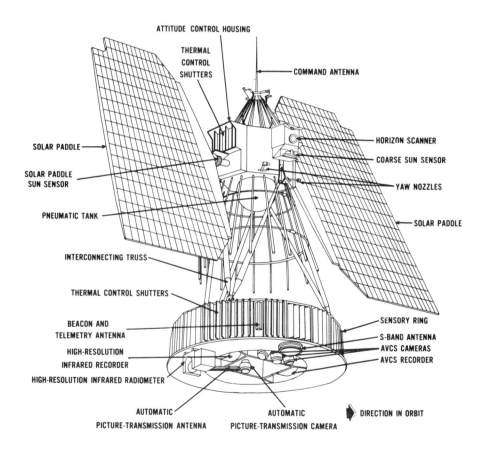

FIG. 2-11. Nimbus meteorological satellite. (Courtesy of NASA. See Appendix 10 for characteristics of meteorological satellites.)

FIG. 2-12. Nomad I weather station. Power is generated by wind and solar batteries. Measurements are made of the wind, the ocean current, air and sea temperature, as well as other atmospheric parameters. (Courtesy of National Bureau of Standards.)

PROBLEMS

Problems marked with an asterisk (°) are the most challenging.

1. Compute the height of a water barometer at a place where the atmosphere's pressure is 850 mb, if the temperature of the barometer is 10°C.

2. The average atmospheric density at sea level is about 1.2×10^{-3} g/cm³, and the average pressure 1013 mb. If the density were constant in the vertical, what would be the depth of the atmosphere (at what point would $p = 0$)? What would be the depth of liquid water having the same pressure (1013 mb) at the bottom?

3. Using the average sea level pressure, compute the total weight of the atmosphere in pounds and kilograms. At 100 miles (160 km) altitude, the atmospheric pressure is only about 0.5×10^{-5} mb (Figure 1-1); how many tons of air remain above this altitude (100 miles)?

4. The dry and wet bulbs of a psychrometer read 60°F and 40°F, respectively. Consulting Appendix 7, determine (a) the dew point; (b) the relative humidity; and (c) the saturation and actual vapor pressures (in. Hg). Convert the units of the two vapor pressures to millibars (Appendix 3) and to density of water vapor (Figure 2-4). How much would the air have to be cooled to cause saturation? How many grams of water would have to be added to each cubic meter of air through evaporation to produce saturation? Which of these two processes leading to saturation is likely to be more quickly accomplished in nature? Explain your answer.

*5. According to Boyle's law (p. 25), at a fixed temperature $p = kd$. Suppose that you have a volume of dry air to which you gradually add water vapor, extracting air molecules so that the pressure and temperature remain unchanged. How will the total mass of gas change? How will the density change? (The "mean" molecular weight of dry air is 29 while that for water is 18.)

6. Make a list of all the factors that you can think of that might determine whether two colliding water drops will coalesce. Can you devise laboratory experiments for testing the relative significance of each factor?

7. If there is a mist in the morning with a temperature of 45°F, what will be the relative humidity in the mid-afternoon, when the temperature is 80°F, assuming that the vapor pressure remains unchanged?

*8. Estimate the average amount of precipitable water over the state of Arizona from Table 2-1 and Figure 2-7. What is the average total volume in cm^3 (or weight in grams) of precipitable water over the state? If the average flow of air across the state is 10 ms^{-1}, how much precipitable water passes over the state in a year? Compare this figure with annual volume of water in the Colorado River. (The average discharge is about 550 m³s^{-1}.) Compare the total annual precipitable water with the actual annual precipitation for the state (Figure 7-10).

*9. How large a volume of air could be saturated by a water droplet of radius 10^{-4} cm if it were completely evaporated? (Assume sea level pressure and temperature of 0°C.) In a typical cumulus cloud over the oceans (p. 39), what proportion of the total water content (liquid + vapor) is in liquid form (sea level pressure, 0°C).

10. The mighty Amazon River carries about 4.8×10^{12} gallons per day. Estimate the area of the "watershed" of this river from a topographic map of South America. Assuming that all the rainfall runs off into the river (neglecting evaporation) what must be the average rainfall for this watershed? How does this value compare with precipitation distribution maps? Look up the average flows of all the other great rivers of the world and sum them; how does the Amazon's flow compare with the combined flows of all other rivers?

11. The evaporation-precipitation cycle behaves like an enormous desalinization plant. If man wanted to double the natural desalinization rate, how much energy would be required each year? How does this compare with the U.S. power production?

3

The Energy of the Atmosphere

3-1.   Energy Budgets and Heat Engines

A convenient way to examine the workings of the atmosphere is through the energy budget. The law of
the conservation of energy requires that we account
for all of the energy received by the earth, so that by
looking at all forms of energy and transformations we
have a guide to atmospheric phenomena. This is similar to following in detail what happens to the fuel
energy provided in an engine, thereby ending up with
a fairly good picture of the operation of the engine.
Figure 3-1 presents a schematic energy flow diagram.
This and the following chapters will deal with particular portions of the flow diagram.

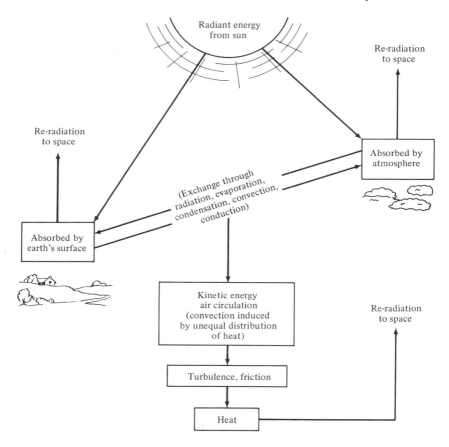

FIG. 3-1.   Energy flow diagram.

Practically all of the energy that reaches the earth comes from the *sun*. Intercepted first by the atmosphere, a small part is directly absorbed, particularly by certain gases such as ozone and water vapor. Some of the energy is reflected back to space by the atmosphere, its clouds, and the earth's surface. Some of the sun's radiant energy is absorbed by the earth's surface. Transfers of energy between the earth's surface and the atmosphere occur in a variety of ways, such as by radiation, conduction, evaporation, and convection. Kinetic energy (air in motion, or wind) results from differences in temperature within the atmosphere, in much the same way that a heat engine converts differences in heat levels between the inside and outside of the expansion chamber to the motion of the piston. And, finally, friction is constantly bleeding off some of the energy of motion, converting it to heat. The combination of these many processes, which are listed in Figure 3-1, produces the complex atmospheric phenomena called weather.

Transfer of Heat Energy

Heat energy can be transmitted from one place to another by conduction, convection, and radiation.

Conduction is the process by which heat energy is transmitted through a substance by point-to-point contact of neighboring molecules, even though the molecules do not leave their mean positions. Solid substances, especially metals, are usually good conductors of heat, but fluids such as air and water are relatively poor conductors. Heat conduction in air is so slow that it is of little importance in transmitting heat within the air itself; however it is significant in the exchange of heat between the earth's surface and the air in contact with it.

Convection transmits heat by transporting groups of molecules from place to place within a substance. Thus, convection occurs in substances in which the molecules are free to move about; i.e., in fluids. The convective motions that carry heat from one point to another within a fluid arise because the warmer portions of the fluid are less dense (recall Charles' law, p. 25) than the surroundings and therefore rise, while the cooler portions are more dense and therefore sink. A circulation of fluid, i.e., a closed circuit of fluid motion, is thus established between the warm and cool regions. Much more will be said about convection in later chapters because it is an important heat transfer process in the atmosphere above the lowest half-meter or so.

Radiation is the transfer of heat energy without the involvement of a physical substance in the transmission. Heat may therefore be transmitted through a vacuum and, if the radiation takes place through a completely transparent medium, the medium itself is not heated or otherwise affected. The transfer of the heat energy from the sun to the earth is by means of the radiative process. Earth also loses its heat energy to outer space in the same way: by radiative transfer.

A simple illustration of the heat transfer processes is given in Figure 3-2. The heat generated by the campfire passes through the pan by conduction, warming the lowest layer of the water in the pan. The rest of the water is heated by convection currents which carry the warmed water throughout the vessel. The camper is warmed mostly by radiation from the fire and the pan, without the intervening air being appreciably

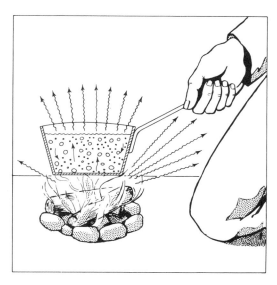

FIG. 3-2. Illustration of conduction, convection, and radiation.

affected. Typically, all three processes go on simultaneously in most situations involving heat transfer.

3-2. Solar Energy

Since the atmospheric "heat engine" is powered by solar energy, we start our discussion with the source of the "fuel"; that is, the sun. The sun is not an unusual star, either in brilliance or size. A slowly rotating body of hot (several million degrees centigrade), very dense gas, with a diameter of about 1,400,000 km (870,000 mi), it is surrounded by a very tenuous atmosphere that extends several solar diameters from the surface (Figure 3-3). The sun generates a tremendous amount of heat (about 5×10^{27} cal are radiated every minute), but the earth intercepts less than one part in two billion of this total. Measurements made on the earth indicate that the rate at which energy impinges on a surface perpendicular to the sun's rays at the mean solar-earth distance is about 2.00 cal/cm²/min.° This value is known as the solar constant, although no one is certain exactly how "constant" is the output of the sun. How-

---

° Recent measurements made at an altitude of 82 km from an X-15 aircraft yielded a value for the solar constant of 1.95 cal/cm²/min.

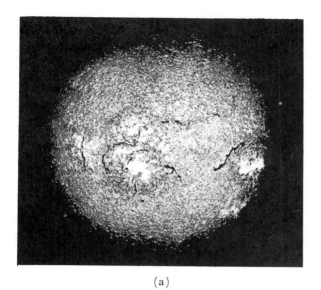

(a)

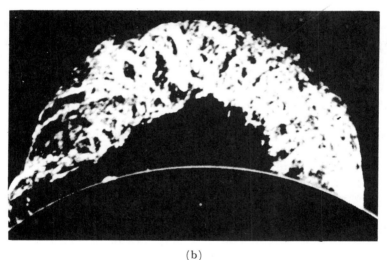

(b)

FIG. 3-3. (a) The solar disk, as seen in the red wavelength of hydrogen. Dark, thread-like features are "filaments," called "prominences" when seen in profile at the edge of the sun. The bright features are plages, and the dark, rather circular ones in the plage regions are sunspots. (Courtesy of Sacramento Peak Observatory, Sunspot, New Mexico.) (b) Largest solar prominence ever recorded. (June 4, 1946. Courtesy of High Altitude Observatory, Climax, Colorado.) (Sun's face has been blacked out by metal disk in coronagraph used to photograph the phenomenon.)

ever, variations of the total energy output are probably smaller than the accuracy of the measurements, which is about $\pm$ 2 per cent.

We know that the sun is not tranquil. For example, it is afflicted by "rashes" on its face [sunspots, Figure 3-3(a)] that have puzzled astronomers since Galileo first noticed them 350 years ago. It is now generally believed that sunspots are essentially spiraling, convective bubbles. The number of spots, as well as their location on the face of the sun, varies quite a bit from time to time. There is a fairly well-defined cycle in the number observed each year, with about 11 years between successive maxima or minima; the number of spots during "minimum" years is usually less than 10 but the number during "maximum" years varies from as few as 43 to as many as 155. There are also intermittent outbursts of small particles and very short wavelengths of radiation associated with disturbances such as those shown in Figure 3-3(b); these have important effects on the upper atmosphere, as was mentioned in Chapter 1.

According to Bethe's theory, the energy radiated from the sun is created through a complex thermonuclear reaction that converts protons (hydrogen nuclei) to alpha particles (helium nuclei). In the process, mass is converted to energy (in accordance with Einstein's familiar relationship, $E = mc^2$). The gravitational contraction of the enormous mass of the sun produces the temperature needed to make such a reaction possible. Although the sun is now converting mass to energy at the rate of about 4×10^6 tons per second, judging from the number of protons still available, the sun should continue shining for another 10^{11} years.

The radiant energy from the sun behaves in part as though it were transmitted in the form of waves, and in part as though it were in the form of particles projected through space. These latter emanations, e.g., cosmic rays, are ordinarily of little consequence in meteorology, and it is usually sufficient to treat the meteorologically significant radiation from the sun as though it were all received in the form of waves. The nature of the radiated energy can be determined by the wavelength (the distance between two successive wave crests or troughs) or by the frequency (the number of times a wave crest or trough passes a given point in a specified unit of time). The *spectrum* of a body is the term for the wavelengths, or frequencies, at which radiated energy is emitted by the body, and the total range of spectra for all such emissions is called the *electromagnetic spectrum*.

The speed of the wave motion is given by the product of the wavelength and the frequency or, in symbolic form,

$$s = L \times f$$

where s is the wave speed, L is the wavelength, and f is the frequency. Since all forms of electromagnetic radiation are propagated at the same speed; i.e., 3×10^{10} cm/sec (186,000 mi/sec), the frequency at any point in the electromagnetic spectrum is inversely proportional to the wavelength.

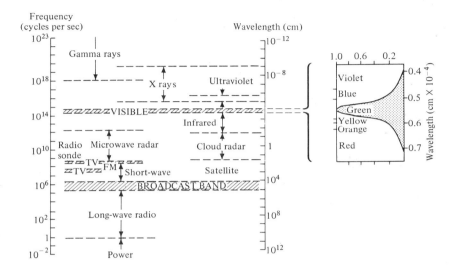

FIG. 3-4.   The electromagnetic spectrum. The curve at the right gives the relative sensitivity of the human eye to electromagnetic radiation (maximum sensitivity at 0.555×10^{-4} cm wavelength.)

The nomenclature for various portions of the spectrum is given in Figure 3-4. It will be noted that only a small portion of the spectrum lies within the range of visible light. Within this narrow band of wavelengths (0.38×10^{-4} to 0.72×10^{-4} cm, approximately) are all of the colors of the rainbow.* Below this range are the ultraviolet, X-ray, and gamma-ray portions of the spectrum; above it are the invisible infrared portions, and the wavelengths associated with radar and radio and television broadcasting.

The amounts of energy emitted at various wavelengths by the sun and by the earth are of significance to the atmospheric scientist because the ability of atmospheric gases and other particles to absorb or to

---

* The wavelengths in the electromagnetic spectrum are frequently given in other units; e.g., microns or Angstrom units. Conversion factors of length units are found in Appendix 3.

deflect incident radiation is dependent on the wavelength. For example, it was pointed out in Chapter 1 that ozone is very effective in absorbing ultraviolet light and that carbon dioxide is an effective absorber in the infrared portion of the spectrum. Both the total rate at which energy is emitted from a body and the distribution of emission rate as a function of wavelength depend on the body's temperature. As expressed by Planck's radiation law, the rate at which a body radiates energy increases with the fourth power of the absolute temperature: αT^4, while the wavelength at which it emits most intensely varies inversely with the temperature: $\alpha(1/T)$. In other words, hot bodies like the sun not only radiate *much more energy* per unit surface area per unit time than cool bodies like the earth, but they do it at *shorter wavelengths*. For example, if the absolute temperature of a body were doubled, its energy radiation rate would increase by 2^4 (16-fold) while the wavelength of its peak emission rate would decrease by half.

The energy spectra predicted by Planck's law at temperatures of 6000°K, 300°K, and 250°K are shown in Figure 3-5. The solar-energy spectrum is very similar to that of the 6000°K-curve of Figure 3-5; in other words, the effective radiative temperature of the sun's surface is approximately 6000°K. The per cent distribution of energy actually observed in various regions of the solar spectrum is given in Table 3-1. It can be seen that much of the solar energy falls in the visible portion of the spectrum with the peak energy occurring at a wavelength of approximately 0.5×10^{-4} cm, which is blue-green. For the earth, on the other hand, with its much lower ($\sim$250°K) effective radiative temperature ("planetary temperature"), the peak radiation occurs at about

TABLE 3-1

Proportion of Total Solar Energy Emitted in Various Wavelength Regions ($1 \text{ cm} = 10^8 \text{ Å} = 10^4 \mu$)

Wavelength	Region	Approximate per cent of total energy
<10 Å	X rays and γ rays	.02
10 Å–2000 Å	Far ultraviolet	
2000 Å–3150 Å	Middle ultraviolet	1.95
3150 Å–3800 Å	Near ultraviolet	5.32
3800 Å–7200 Å	Visible	43.50
7200 Å–1.5 μ	Near infrared	36.80
1.5 μ–5.6 μ	Middle infrared	12.00
5.6 μ–1000 μ	Far infrared	0.41
>1000 μ	Micro- and radio waves	

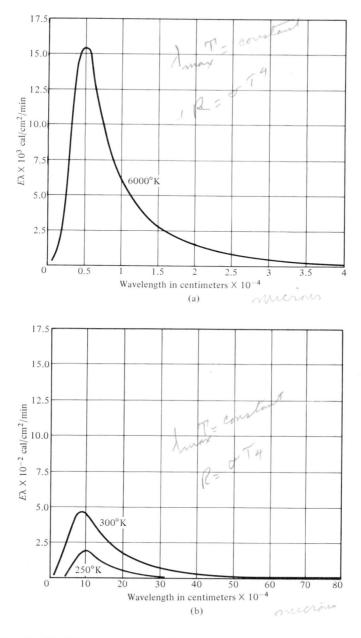

FIG. 3-5. Black-body emission of (a) a hot body such as the sun, and (b) a cool body such as the earth. [In comparing the curves, note that the vertical scale of (a) is 100,000 times that of (b).]

67

10^{-3} cm, well into the invisible infrared portion. The *total* energy in all wavelengths emitted by the earth is about 1/160,000 of that emitted by the sun.

What happens to the enormous amount of energy which reaches the outer fringes of the earth's atmosphere in the form of electromagnetic waves? As these rays encounter the atmosphere, some are absorbed by the atmosphere, some are reflected back to space, and some are scattered within the atmosphere and are eventually absorbed or lost again to space. The remainder of the radiated energy passes through the atmosphere undisturbed and is intercepted by the earth itself. We shall examine in some detail how these various processes take place.

Absorption by the Atmosphere

Oxygen, ozone, water vapor, carbon dioxide, and dust particles are the most significant absorbers of both the "short-wave" radiation from the sun and the "long-wave" radiation from the earth. However, the gases are *selective* absorbers, meaning that they absorb strongly in some wavelengths, less strongly in others, and hardly at all in still others.

Figure 3-6 summarizes the relative effectiveness of the principal absorbing gases of the atmosphere. (Absorptivity is the fractional part of incident radiation that is absorbed.) The very short ultraviolet (less than 0.20×10^{-4} cm) radiation of the sun is absorbed as it encounters and splits molecular oxygen into two atoms in the upper levels of the atmosphere (Chapter 1), while the ultraviolet light of larger wavelengths (specifically, that between 0.22×10^{-4} and 0.29×10^{-4} cm) is very effectively absorbed by ozone (O_3). As can be seen from Figure 3-6, oxygen (O_2) and ozone (O_3) absorb almost 100 per cent of all radiation at wavelengths less than 0.29×10^{-4} cm. For this reason, only a minute portion of the sun's ultraviolet radiation penetrates to the lower levels of the atmosphere. Because excessive exposure to ultraviolet radiation is harmful to human beings and many other forms of life, this absorption by oxygen and ozone provides a very fortunate protective shield against such radiation. In the longer wavelengths, neither oxygen nor ozone absorbs very much energy, except for a narrow band near 9.6×10^{-4} cm, in the infrared region. About 2 per cent of the sun's total radiation received at the outer fringes of the atmosphere is depleted by ozone.

Water vapor is a significant absorber of radiation. Its complicated absorptivity characteristics are illustrated in Figure 3-6(b). Although not effective at wavelengths below 0.8×10^{-4} cm, where most of the *solar* radiation occurs, it absorbs strongly in several narrow zones in the infrared region, in fairly broad zones (2.5×10^{-4} to 3.5×10^{-4} and 5×10^{-4} to 7×10^{-4} cm), and moderately well beyond 13×10^{-4} cm.

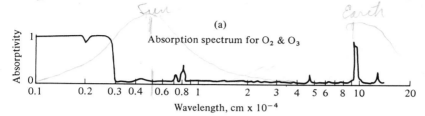

(a)
Absorption spectrum for O_2 & O_3

Wavelength, cm x 10^{-4}

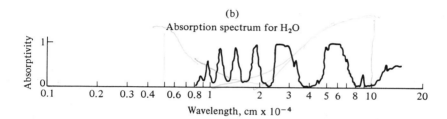

(b)
Absorption spectrum for H_2O

Wavelength, cm x 10^{-4}

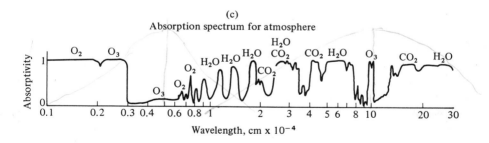

(c)
Absorption spectrum for atmosphere

Wavelength, cm x 10^{-4}

FIG. 3-6. Absorption of radiation at various wavelengths by (a) O_2 and O_3; (b) H_2O; and (c) the principal absorbing gases.

These are wavelengths at which the relatively cool earth and its atmosphere emit much of their energy (Figure 3-5).

Carbon dioxide has an absorption spectrum that overlaps that of water vapor to some extent. It too absorbs mostly in the infrared region in various narrow zones of moderate effectiveness (1.6×10^{-4},

2.0×10^{-4}, 2.7×10^{-4}, and 4.3×10^{-4} cm), and in a strong band near 15×10^{-4} cm.

In summary, the atmosphere is essentially transparent between 0.3×10^{-4} and 0.8×10^{-4} cm, where most of the solar (short-wave) radiation occurs (Figure 3-5 and Table 3-1). But between 0.8×10^{-4} and 20×10^{-4} cm, where much of the terrestrial radiation is emitted, there are several bands of moderate absorptivity which are the result of the absorbing characteristics of water vapor and carbon dioxide. Note, however, that there is a significant transparent region in the infrared (8×10^{-4} to 12×10^{-4} cm) which meteorologists refer to as the "atmospheric window"; it encompasses the point where the peak earth radiation occurs.

Scattering

The atmosphere is composed of very many discrete particles (gas molecules, dust, water droplets, etc.) but the empty space between particles is actually much greater than the volume occupied by the particles. Each particle acts as an obstacle in the path of radiant energy, such as light waves, traveling through the atmosphere, much as rocks in a lake impede the progress of ripples in the water. The waves are deformed by these obstacles into a pattern so that the rays appear to emanate from the obstacles. Thus, radiant energy propagating in a single direction is dispersed in all directions as it encounters each particle in its path. This dispersion of the energy is called scattering.

The nature of the scattering depends partly on the size of the particles and partly on the wavelength of the radiation. For particles the size of gas molecules, the amount of scattering is much higher for the short wavelengths of light (blue) than for the long wavelengths (red), as illustrated by Figure 3-7 and Table 3-2. It is for this reason that the sky appears blue—it derives its color from the short waves that are readily scattered by air molecules. In the higher atmosphere, the density of the scattering particles is lower; therefore, the amount of scattering is decreased. Thus, astronauts have observed, as they ascend through the atmosphere, that the sky becomes darker and finally becomes black as the effect of scattering is eliminated. When the sun or moon is near the horizon, the bluish colors from the visible spectrum have been removed by scattering, and the remaining light appears yellow or red— thus providing the beautiful colors of sunrise or sunset.

When the atmosphere contains many large dust particles or minute water droplets (haze), scattering is no longer very selective in terms of wavelength. The long waves are scattered almost as much as the short

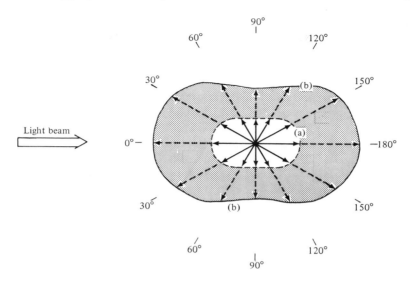

FIG. 3-7. The relative intensities of the light scattered in all directions by a small particle for (a) red and (b) green. (Intensities are proportional to arrow lengths. Note that amounts scattered forward and backward are double the amounts scattered at right angles to the beam.)

waves, and the resulting sky color becomes less bluish and more white or milky. For this reason, the blueness of a cloudless sky is an indication of its "purity"; i.e., how free it is of smoke, dust, and haze.

It is evident from Table 3-2 that the depletion of ultraviolet light (wavelengths less than 0.3 μ) by scattering is so great that even without the strong absorption by O_2 and O_3, very little energy in these wavelengths would reach the surface of the earth.

TABLE 3-2

Per Cent Radiation Reaching Ground After Scattering by Air Molecules of the Atmosphere

Wavelength (μ):	0.20	0.25	0.30	0.35	0.40	0.45	0.50	0.60
% Transmitted:	0.05	6.7	29.5	53.0	69.6	80.0	86.5	93.3

Wavelength (μ):	0.70	0.80	0.90	1.00	1.50	2.00	4.00
% Transmitted:	96.4	97.9	98.7	99.1	99.8	99.9	100.0

Reflection

The radiant energy from the sun encounters still another obstacle before it reaches the surface of the earth—clouds. Most clouds are very good reflectors, poor absorbers, of radiant energy. The reflection by clouds depends primarily on their thickness, but also to some extent on the nature of the cloud particles (i.e., whether ice or water), and the size of these particles. For water-droplet clouds containing one gram per cubic meter of liquid water and having an average drop diameter of 0.001 cm, Figure 3-8 shows the relationship between cloud thickness and the percentage reflection by clouds of varying thicknesses. The amount of energy transmitted through the cloud, as well as that absorbed by the cloud, is also shown. It will be noted that the reflection by clouds varies from less than 50 per cent to more than 90 per cent, depending on the cloud thickness. It will also be observed that a thin cloud deck will transmit a significant portion of the radiated energy directly through the cloud, and that regardless of thickness, clouds absorb very little radiated energy.

In general, the earth's surface is a poor reflector of solar radiation, although the amount of reflection varies greatly with the nature of the surface. Table 3-3 shows the reflectivity for a number of common sur-

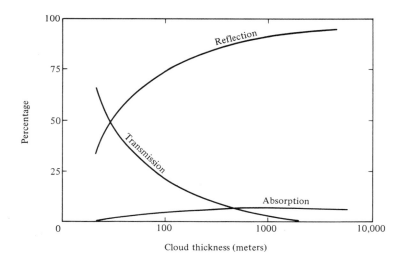

FIG. 3-8. Percentage reflection, absorption, and transmission of solar radiation by clouds. (After Hewson, *QJRMS*.)

TABLE 3-3

Reflectivity or "Albedo" of Various Surfaces

Surface	Per cent reflected
Clouds (stratus) <500 ft thick	5–63
500–1000 ft thick	31–75
1000–2000 ft thick	59–84
(average of all types and thicknesses)	50–55
Concrete	17–27
Crops, green	5–15
Forest, green	5–10
Meadows, green	10–20
Ploughed field, moist	14–17
Road, blacktop	5–10
Sand, white	34–40
Snow, fresh fallen	75–90
Snow, old	45–70
Soil, dark	5–15
Soil, light (or desert)	25–30
Water	8*

* Typical value for water surface, but the reflectivity increases sharply from less than 5 per cent when the sun's altitude above the horizon is greater than 30°, to more than 90 per cent when the altitude is less than 3°. The roughness of the sea surface also affects the albedo somewhat.

faces. It will be noted that the reflection from a thick cloud layer is significantly greater than from a thin layer, while on the solid earth, light-colored objects reflect more than dark-colored objects.

Absorption by the Earth

Figure 3-9 presents a summary of what happens, on the average, to the solar radiation intercepted by the earth and its atmosphere. Although normally about 50 per cent of the earth's surface is covered by clouds, they absorb only about 2 per cent of the short-wave radiation. Approximately 17 per cent is absorbed by the gases and dust of the atmosphere, principally by water vapor. This means that a total of 19 per cent is absorbed by the atmosphere and its constituents.

The earth's surface absorbs about 47 per cent of the solar radiation. Some of it (19 per cent) comes directly from the sun; some (23 per cent) after reflection by clouds; the rest (5 per cent) is received after being scattered by the air. Thus, of the total energy arriving from the sun, approximately 66 per cent is absorbed by the earth's surface and atmo-

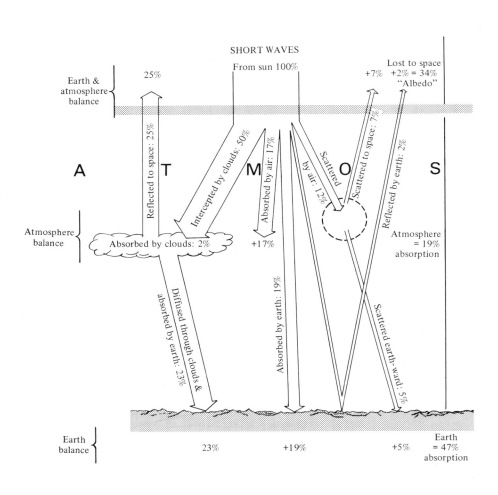

S   P   A

SHORT WAVES

From sun 100%

Lost to space
+7% +2% = 34%
"Albedo"

25%

Earth &
atmosphere
balance

Reflected to space: 25%

Intercepted by clouds: 50%

Absorbed by air: 17%

Scattered by air: 12%

Scattered to space: 7%

Reflected by earth: 2%

A   T   M   O   S

Atmosphere
balance

Absorbed by clouds: 2%

+17%

Atmosphere
= 19%
absorption

Diffused through clouds &
absorbed by earth: 23%

Absorbed by earth: 19%

Scattered earth-ward: 5%

Earth
balance

23%

+19%

+5%

Earth
= 47%
absorption

E   A   R

FIG. 3-9. The earth's heat balance.

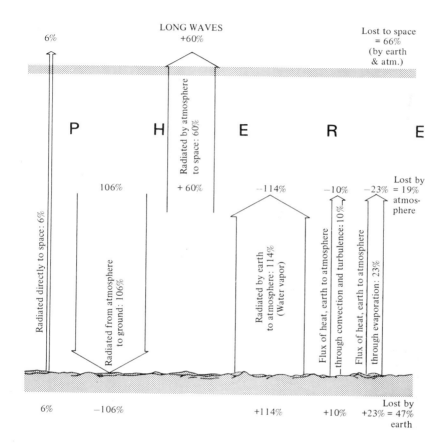

C   E

6%   LONG WAVES   Lost to space
              +60%   = 66%
                                (by earth
                                & atm.)

P   H   E   R   E

Radiated by atmosphere to space: 60%

106%   +60%   −114%   −10%   −23%   Lost by
                                            = 19%
                                            atmos-
                                            phere

Radiated directly to space: 6%

Radiated from atmosphere to ground: 106%

Radiated by earth to atmosphere: 114% (Water vapor)

Flux of heat, earth to atmosphere — through convection and turbulence: 10%

Flux of heat, earth to atmosphere through evaporation: 23%

6%   −106%   +114%   +10%   +23% = 47%   earth
                                              Lost by

T   H

sphere. The rest, 34 per cent, is lost to space, having been reflected by the clouds and the earth's surface or scattered by the particles in the air. This average reflectivity or *albedo* of the earth (0.34) represents the fractional part of the incident radiation that is bounced off the earth. For purposes of comparison, the moon's albedo is only about 7 per cent, which means that it is not nearly as bright as the earth.

3-3. The Earth's Heat Balance

Throughout geologic history, the appearance of ice ages in middle latitudes suggests that major changes in the earth's climate have occurred. Such changes may have been associated with long-period imbalances in the amount of incoming and outgoing radiation on the earth. However, over moderately long periods of time (between hundreds and thousands of years), studies have shown that the average temperature of the earth remains essentially constant. This would indicate that there exists a long-term heat balance between the earth and space. It follows, then, that of the total amount of solar energy absorbed by the earth and its atmosphere (66 per cent), an equal amount must be reradiated back to space.

The manner in which this "heat budget" is balanced, and the approximate amounts reflected, scattered, absorbed, and reradiated, are shown in Table 3-4. Here, of the average amount of short-wave solar radiation received by the earth and its atmosphere (0.50 cal/cm²/min),* 0.33 cal/cm²/min is absorbed by the atmosphere and the earth. The remainder (0.17 cal/cm²/min) is immediately lost to space by reflection and scattering. After going through the complex exchange processes described in Figures 3-1 and 3-9, energy which was absorbed is reradiated again to outer space, thus keeping the budget in balance.

It should be noted that, although there is a heat balance for the planet as a whole, all parts of the earth and its atmosphere are not in radiative balance. In fact, it is the imbalance between the heat surplus received at the equator and the deficit at the poles that drives the atmospheric "heat engine." The transfer of heat energy between these warm and cold regions of the earth produces the global wind system. The winds, in turn, reflect the work done by the atmospheric engine.

Of special interest on the long-wave (right-hand) side of Figure 3-9 and in item III of Table 3-4 is the fact that the amounts of radiative energy emitted by the air to the earth (106 per cent or 0.53 cal cm^{-2}

---

* See Problem 2 at end of this chapter.

TABLE 3-4

The Energy Budget of the Atmosphere (See Also Figure 3-9)

I. INCOME—Short-Wave Radiation (cal/cm²/min)
    Average solar radiation received by earth and atmosphere: 0.50
        Absorbed by: (a) atmosphere and clouds   0.09   *19 %*
                     (b) earth's surface   0.24   *47 %*
                     Net short-wave absorbed   0.33   *66 %*

II. OUTGO—Short-Wave Radiation (cal/cm²/min)
     Lost to space by: (a) reflection from clouds   0.13   *25 %*
                       (b) reflection from earth   0.01   *2 %*
                       (c) scattering from air   0.03   *7 %*
        (i) Net short-wave outgo   0.17   *34 %*

III. OUTGO—Long-Wave Radiation (cal/cm²/min)
      The atmosphere: (a) radiates to the earth   0.53   *106 %*
                      (b) radiates to space   0.30   *60 %*
                      (c) receives from earth through
                          radiation   −0.57°   *114 %*
                      (d) receives from earth through
                          convection, turbulence, and
                          evaporation   −0.17   *33 %*
         (ii) Net long-wave outgo from atmosphere   0.09   *19 %*
      The earth: (a) radiates to space   0.03   *6 %*
                 (b) radiates to the atmosphere   0.57   *114 %*
                 (c) loses to the atmosphere through convection
                     and turbulence   0.05   *10 %*
                 (d) loses to the atmosphere through evaporation 0.12   *23 %*
                 (e) receives from the atmosphere through
                     radiation   −0.53   *106 %*
         (iii) Net long-wave outgo from earth   0.24   *47 %*
             Total net outgo (i) + (ii) + (iii):   0.50

° Negative signs under "outgo" indicate net gains.

min⁻¹) and by the earth to the air (114 per cent or 0.57 cal cm⁻² min⁻¹) actually exceed the total solar energy absorbed by the earth. This is due to the "blanketing" effect of the atmosphere, which keeps the earth's surface and lower layers of the atmosphere a good deal warmer than they would be without the atmosphere. On the moon's sunlit surface, which absorbs twice as much solar energy per unit area as does the earth's surface, the average temperature is more than 20°C colder than that of the earth because the moon lacks an atmosphere.

Two gases—water vapor and carbon dioxide—play the most important role in this function of keeping the earth warm. Except for the "window" between about 8×10^{-4} and 12×10^{-4} cm [Figure 3-6(c)], these gases block the direct escape of the infrared energy emitted by the earth's surface. Although the atmosphere absorbs but a small percentage of the short-wave solar radiation, it is quite opaque to the long-wave terrestrial radiation. Only when the earth's surface temperature is fairly high does the radiational loss through the transparent bands and from the top of the atmosphere equal the amount absorbed from the sun. This heat-retaining behavior of the atmosphere is analogous to what happens in a greenhouse, where the glass (or more recently, plastic) roof and sides trap the sun's heat energy to keep the greenhouse warm. For this reason, the effect of moisture in the atmosphere in trapping the solar energy is called the *greenhouse effect*. This effect is quite noticeable when one compares the rapid temperature fall at night in the desert, where the air is dry, with the slower decrease in temperature in coastal regions, where the air is moist. On a global basis, the effectiveness of the atmospheric "greenhouse" is illustrated by the fact that the mean temperature near the surface is about 15°C while the overall "planetary temperature" is only about −25°C (p. 66).

3-4. Distribution of the Earth's Heat Energy

It has already been mentioned that, considering the earth as a whole, the incoming and outgoing radiant energy is essentially in balance. However, it is not in balance everywhere on earth. This is primarily because the amount of incoming solar energy absorbed at the earth's surface varies greatly with the earth's motions, the distribution of physical properties of the surface, and the nature and amount of cloudiness. To a lesser extent, the amount of energy *emitted* by the earth and its atmosphere to space varies because of differences in the moisture contained in the air.

Variability of Incoming Solar Energy

The latitudinal variations of incoming solar energy are related to three characteristics of the earth and the sun: (1) the earth is essentially a sphere, (2) the sun is so far away that its rays of radiated energy are approximately parallel, and (3) the earth is rotating. Only one-half of the earth can be illuminated at one time, and the angle that the sun's rays make with the sphere's surface will decrease from 90° at the exact center of the illuminated circle to zero degrees at the edges where the shadow begins. This is illustrated in Figure 3-10(a). Angle *c*, at the

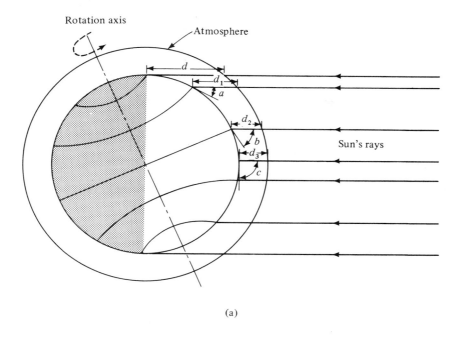

(a)

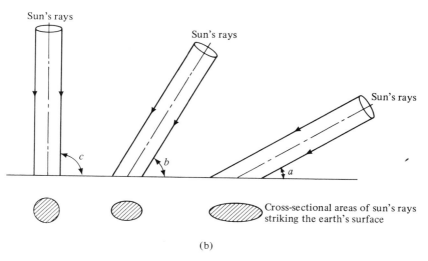

Cross-sectional areas of sun's rays striking the earth's surface

(b)

FIG. 3-10. The intensity of solar radiation depends on the angle at which the sun's rays strike the earth's surface. (a) The angles of incidence, a, b, c, and the depths of penetration through the atmosphere, d_1, d_2, d_3, at different parallels of latitude; (b) The varying cross-sectional area on the earth's surface due to different angles of incidence.

center of the illuminated circle, is 90° since this is the one point where the sun's rays are perpendicular to the surface. Angle b is clearly less than 90°, and angle a is smaller than angle b. The rays entering with angle a will, of course, traverse a greater mass of atmosphere than those entering with angle b or c (since the distance $d_1 > d_2 > d_3$). For example, the mass of air that must be traversed by a beam entering at an angle of 60° from the vertical is twice that for one coming in from the zenith; and when the sun is only 5° above the horizon, the rays must pass through a mass more than 10 times greater than they would if they were coming from overhead. The rays entering at angle a will therefore be subject to greater depletion by absorption, reflection, and scattering. But even more important, the intensity will be smaller at lower solar angles because the same amount of energy will intercept larger surface areas, as illustrated in Figure 3-10(b). Thus, the amount of energy received by each unit of surface area will be smaller when the sun is near the horizon than when it is overhead.

Finally, the earth rotates about an axis, and the axis of rotation is not perpendicular to the plane containing the path of the earth's revolution around the sun, but rather it is tilted at an angle of 23½ degrees from the perpendicular. Consequently, the half of the sphere that is illuminated by the sun is continuously changing.

Consider the ring formed by the circular edge of the earth's shadow (Figure 3-11). When the axis of rotation lies in the same plane as this

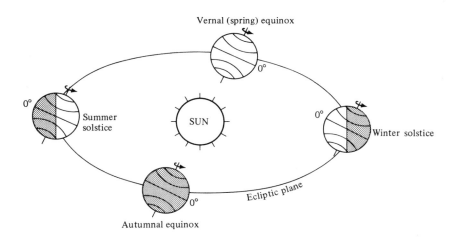

FIG. 3-11. The revolution of the earth about the sun, illustrating the changing length of day.

ring, it is evident that the edge of the shadow will divide each latitude circle exactly in half, so that all places on earth will have precisely 12 hours of sunlight and 12 hours of darkness. This occurs at the time of the spring and fall equinoxes (March 21 and September 23, approximately). But when the axis of rotation does not fall in the plane of the shadow's ring, the latitude circles will not be divided into two equal parts, except at the equator. Thus, the ratio of night to day at each latitude will be in the same proportion as the two segments of the latitude circle created by the shadow ring. Table 3-5 gives the length of day as a function of latitude at the solstices and equinoxes.

TABLE 3-5

Length of Day

Latitude	Vernal equinox	Summer solstice		Autumnal equinox	Winter solstice	
0°	12 hr	12 hr		12 hr	12 hr	
10°	12	12 hr	35 min	12	11 hr	25 min
20°	12	13	12	12	10	48
30°	12	13	56	12	10	4
40°	12	14	52	12	9	8
50°	12	16	18	12	7	42
60°	12	18	27	12	5	33
70°	12	2 mo		12	0	0
80°	12	4 mo		12	0	0
90°	12	6 mo		12	0	0

Figure 3-12 shows the way in which the total energy received each day *(insolation)* varies with the time of year and latitude. During the summer in both hemispheres, the total daily energy received varies little between the poles and the equator because the lower solar angles in the polar regions are compensated for by the greater duration of sunshine each day. In the winter, of course, the latitudinal variation is very great since little or no energy is received at high latitudes, while that received near the equator remains almost unchanged throughout the year.

It may also be noted from Figure 3-12 that the Southern Hemisphere receives a little more energy during its summer than does the Northern Hemisphere in its summer, and conversely during their respective winters. These small differences are due to the fact that the earth moves in a slightly elliptical path about the sun; at the beginning of January the sun and earth are closest together *(perihelion)*, and at

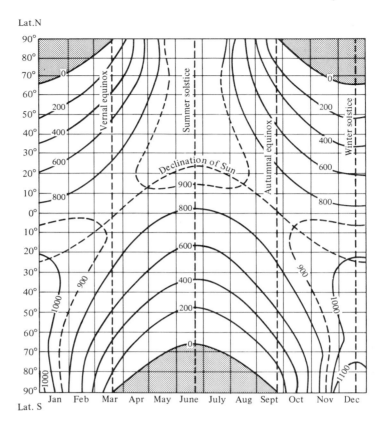

FIG. 3-12.   Undepleted insolation, in cal/cm²/day, as a function of latitude and date. Shaded areas represent latitudes within the earth's shadow.

the beginning of July they are farthest apart (*aphelion*). The difference in total energy received by the earth between aphelion and perihelion is about 7 per cent.

The amount of incoming and outgoing energy, averaged over the entire year, is shown for each latitude in Figure 3-13. It can be seen that, as would be expected, the incoming short-wave energy decreases a great deal between the equator and the poles, while the outgoing long-wave energy is nearly constant.* Although the *total* incoming

---

* Recent satellite measurements of the outgoing energy are presented in Figure 3-14.

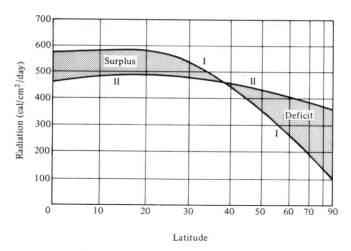

FIG. 3-13. Mean annual insolation (curve I) and outgoing long-wave flux (curve II), at the tropopause.

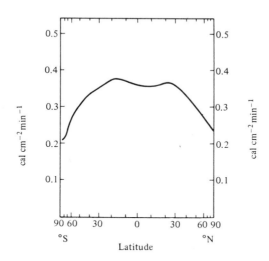

FIG. 3-14. Infrared radiation leaving the earth, averaged along lati-tudes, as measured recently from meteorological satellites. Note that the average emission for the entire earth is close to the previously determined value of 0.33 cal cm^{-2} min^{-1} (Table 3-4).

energy for all latitudes (curve I) equals the *total* outgoing energy for
all latitudes (curve II), income equals outgo only at a single latitude:
near 38 degrees. There is a surplus of incoming energy at tropical
latitudes, and a deficit in the polar regions. If the global wind systems
(and, to a lesser extent, ocean currents) did not exist to redistribute the
energy, the poles would become steadily colder and the tropics steadily
warmer.

Heat Energy and Temperature

If there were only latitudinal variations in the energy received and
absorbed by the earth and its atmosphere, meteorology would be a
considerably simpler study. However, the absorptive properties of the
air and the surface of the earth are not distributed in a smooth, unchanging pattern. The ability of the atmosphere to absorb radiative
energy depends on the clouds, moisture, and dust in the air, and the
concentrations of these may vary both geographically and with time.
The nature of the earth's surface and its absorptive properties are also
distributed erratically over the globe, and even they change with time.
The variation in the ability of various surfaces to absorb radiant energy
is demonstrated in Table 3-3.

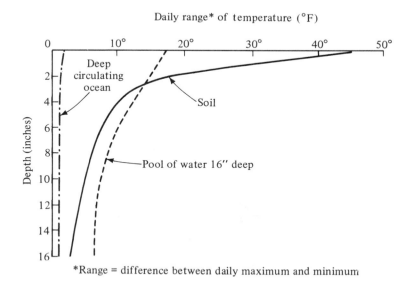

*Range = difference between daily maximum and minimum

FIG. 3-15. Examples of daily temperature ranges at various depths
on a clear summer day in soil, shallow water, and the ocean.

The most pronounced differences in surface thermal properties are those between land and bodies of water. Under identical insolation conditions (same solar angles, duration of daylight, and atmospheric transparency), the temperature changes experienced by the water will be much less than those of the land surface. (See Figure 3-15.) Primarily, this is because water is fluid and so can be mixed both horizontally and vertically. As a result, its heat tends to be distributed over a much greater mass than is the case with the "stagnant" land. The heat absorbed by a land surface is generally confined to the upper few inches, while in water the heat may be distributed to depths of hundreds of feet. Other reasons for the smaller temperature range of water surfaces are: (1) Because water is transparent, radiation can penetrate to significant depths, so that the energy is absorbed by a larger mass of water than is the case for land (see Table 3-6); (2) Water generally has a higher specific heat than does land (i.e., more heat is required to raise the temperature of a gram of water 1°C than is required to produce the same temperature change in land); (3) More evaporation generally takes place from water surfaces than from land, so that the latent heat of evaporation extracted from the water is greater than that taken from the land.

The fact that the oceans act as heat reservoirs is illustrated by the January and July mean air temperature maps of Figure 3-16. It will be observed that the temperature variation between seasons over middle and high latitudes is much greater over the continents than

TABLE 3-6

Penetration of Solar Radiation into Soil and Water (Per Cent° of Incident Energy Remaining at Each Depth)

Depth	Pure water Visible light	Infrared light	Sand or soil (grain size, 0.5–1.0 mm)
0.1 mm	100	100	
0.5	100	100	72
1.0	100	100	54
1.5	100	99	32
2.0	100	99	6
3.0	100	99	1
5.0	100	98	0
10.0	100	98	
100.0	98	85	
1.0 m	97	36	
10.0 m	73	3	
100.0 m	6	0	

° Rounded to nearest whole number.

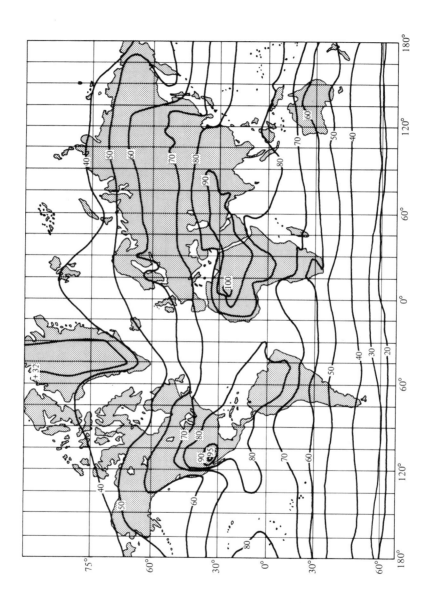

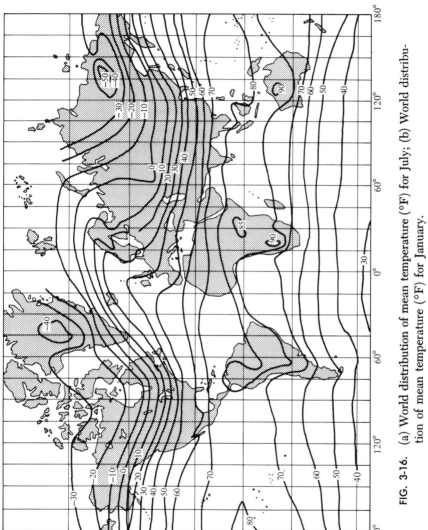

FIG. 3-16. (a) World distribution of mean temperature (°F) for July; (b) World distribution of mean temperature (°F) for January.

over the oceans. For example, at latitude 45°N, the annual range of temperature over the continents is about 60°F, while over the Pacific Ocean at the same latitude, the range is only about 10°F. It will also be noted that the isotherms (lines of equal temperature) dip equatorward over the oceans in summer and poleward in winter, indicating that the ocean is cooler than the land in summer and warmer than the land in winter.

Temperature Lag

The times of high and low air temperature do not coincide with the times of maximum and minimum solar radiation, either on an annual

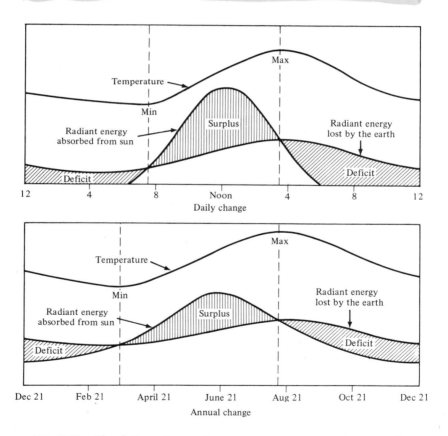

FIG. 3-17. The daily and annual temperature maxima and minima lag behind the maxima and minima of solar radiation.

or daily basis. The months of July and August are generally the warmest of the year, while January and February are the coldest. Yet, on an annual basis, the greatest intensity of radiation occurs in June and the lowest in December (Northern Hemisphere). Furthermore, the highest temperature of the day normally occurs about 3 or 4 P.M. Yet, the greatest intensity of solar insolation each day occurs near noon.

This *lag* in the temperature can be explained on the basis of the time required for heating and cooling (Figure 3-17). The earth loses heat continuously through radiation to outer space. During some months of the year and some hours of the day, the incoming energy from the sun exceeds the outgoing energy lost by the earth. When this is occurring, the temperature will rise, since the heat content of the air is increasing. Thus, the maximum temperature will be recorded at the time when the incoming energy ceases to exceed the outgoing. Thereafter, when the outgoing energy is greater than the incoming, the temperature will fall until the two are again in balance, at which time the lowest temperature will occur.

PROBLEMS

Problems marked with an asterisk (*) are the most challenging.

1. Compute the total solar energy per minute intercepted by the earth, and the fraction of the total energy output of the sun which this represents. (Hint: Recalling the definition of the solar constant, compute the *cross-sectional* area of the earth and the surface area of an imaginary sphere surrounding the sun at the mean solar-earth distance; form the ratio of the two areas.)

*2. Compute the average rate of energy per unit area intercepted by the earth. (Hint: The solar constant, 2.00 cal cm^{-2} min^{-1}, represents the energy per minute striking a surface *perpendicular* to the solar rays; the total energy intercepted by the earth would depend on the cross-sectional area of the earth, but this total energy is distributed, on the average, over the entire surface area of the earth.)

3. Suppose you were determining the temperature distribution on the surface of the moon by measuring the infrared radiation. Considering the transparency of the atmosphere, what wavelength band would give the best results?

4. Compute the elevation angle of the sun at noon at the latitude of your city at the times of the equinoxes. What will be the length of daylight (not counting twilight)?

5. What is the essential difference between the phrases "light from the sun" and "radiation from the sun"?

6. What angle between the earth's axis of rotation and the plane containing the earth's path around the sun would provide (a) the least difference between seasons? (b) the greatest difference?

7. The earth is closer to the sun during January than during July. Yet you will note from Figure 3-16 that July (summer) temperatures in the Northern Hemisphere are generally higher than January (summer) temperatures in the Southern Hemisphere. Why?

8. Explain why nighttime temperatures are generally lower on nights when the humidity is low than when it is high.

9. Suppose you wish to design your living room so that it receives as much sunshine as possible during the winter, and as little as possible during the summer. If you live in the Northern Hemisphere, in which direction should the windows face? What if you live in the Southern Hemisphere?

°10. According to p. 46, 1.4×10^{17} gallons of water are evaporated annually from the earth's surface into the atmosphere. How much heat is used for this purpose each year? What percentage of the total solar energy absorbed by the earth does this represent? Compare your answers with the values given in Table 3-4 and Figure 3-9.

11. How do you explain the difference in albedo between the earth and the moon? Is the dark side of the moon illuminated by "earthlight" as we are by moonlight?

4

Atmospheric Motions: Causes

4-1. The Nature of Air in Motion

As we have already seen, the energy radiated from the sun is distributed unevenly over the earth's surface, partly because of latitudinal variations in the solar insolation, and partly because of differences in the absorptivity of the surface of the earth and its atmosphere. This uneven distribution of heat provides the basis for the conversion of thermal energy into the energy of motion. The mechanics of this conversion of heat energy to kinetic energy is the subject of this chapter.

In its movements through space, the earth is accompanied by its atmosphere. The atmosphere also rotates with the earth from west to east, so that at the equator, the air moves eastward at a speed of more than 450 meters per second (1000 mph); at latitude 60° it moves eastward at half that speed and, of course, at the poles its eastward speed is zero. However, because the ground moves eastward at these same speeds, this rotational motion of the air is not apparent to the earthbound observer. We shall therefore be concerned only with the *deviations* of air motion from those that are due to the motions of the earth itself—such motions of the air *relative* to the earth are what we know as the winds.

4-2. The Forces That Produce Motion

According to the first law of Isaac Newton, a body will *change* its velocity of motion only if acted upon by an unbalanced force. Newton's second law goes on to state that the acceleration (the rapidity of a velocity change) is directly proportional to the magnitude of the un-balanced or "net" force and inversely proportional to the mass of the body. The body will accelerate in the direction of the unbalanced force. In symbolic terms, the second law may be written:

$$\mathbf{A} = \frac{\Sigma\mathbf{F}}{M}$$

where $\Sigma\mathbf{F}$ is the sum of all the forces (therefore, the *net* force), M is mass, and $\mathbf{A}$ is the acceleration.* Thus, to determine the state of motion of a body, one must identify all of the forces that act on the body. In the case of the atmosphere, the following are the most significant forces (per unit mass) that affect air motion:

$\mathbf{P} =$ force produced by pressure;

$\mathbf{G} =$ force produced by gravitational attraction;

$\mathbf{C} =$ deflective force due to the earth's rotation;

$\mathbf{F} =$ force due to friction or "drag."

Thus, for the atmosphere, the equation governing atmospheric motion may be written:

$$\mathbf{A} = \mathbf{P} + \mathbf{G} + \mathbf{C} + \mathbf{F} \qquad (4\text{-}1)$$

---

* Boldfaced letters designate quantities that have *both* magnitude and direction (vectors). When the same letters do not appear in boldface, it means that only the magnitudes are being considered.

Of the four forces in the above equation, the pressure and gravity forces can be thought of as the fundamental or basic forces, since without them there would be no motion. They are the "driving" forces, and are independent of the velocity; i.e., they exist regardless of the state of the motion of the air. In contrast, the Coriolis and frictional forces are *dependent* on the velocity; i.e., they arise *only* when the air is moving. We shall examine each class of atmospheric forces separately.

Forces Independent of Velocity

1. *Gravity.* The gravitational pull of the earth is always directed downward, toward the center of the earth. The strength with which it acts on any "parcel" or "element" of fluid is proportional to the mass of the parcel. It is common in physics to speak in terms of the force per unit mass of this gravitational attraction, which we refer to as the "acceleration of gravity" or merely "gravity." Although gravity varies slightly with latitude and altitude, and even with the irregular mass distribution within the earth, for most meteorological purposes the following mean sea level value may be used:

$$G = 980.6 \text{ cm sec}^{-2} \qquad (4\text{-}2)$$

2. *Force due to pressure.* As we mentioned in Chapter 2, the atmospheric pressure at any point is due to the weight of the air pressing down from above; that is, it results from the force of gravitational attraction. Because of the properties of fluids, the resulting pressure at such a point acts equally in all directions. Thus, in order for a "parcel" or "element" of fluid to experience a net force due to pressure in some particular direction, one "side" of the parcel must be acted on by a pressure different from that acting on the opposite "side." For example, in Figure 4-1, the fluid in the pipe will remain stationary if the pressure p_1 ($= F_1/A$) at the face of the piston on the right is equal to the pres-

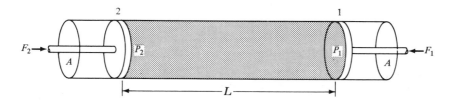

FIG. 4-1. Pressure gradient in a pipe. (A = cross-sectional area of pipe; L = distance between pistons.)

sure p_2 ($= F_2/A$) at the piston face on the left. Only if the forces F_1 and F_2 (and thus the two pressures, p_1 and p_2) *differ*, will the fluid accelerate.

If we assume that the pipe has a cross-sectional area, A, the volume of fluid upon which the force is acting is given by $A \times L$, where L is the distance between the pistons. If d is the density of the fluid, the mass of fluid between the pistons is volume $\times$ density $= A \times L \times d$. Therefore, the net force per unit mass acting on the fluid between the pistons is

$$\frac{F_1 - F_2}{A \times L \times d} = \frac{p_1 - p_2}{d \times L} \tag{4-3}$$

Thus, the force per unit mass in a fluid is directly proportional to the pressure differences $(p_1 - p_2)$, and inversely proportional to the distance (L) between the boundaries of the fluid upon which the pressure forces are acting. This ratio of pressure difference to distance, $(p_1 - p_2)/L$, is called the *pressure gradient*, since it measures the grade or slope of the pressure. It is important to note that the direction of the pressure gradient, and thus also the acceleration it will produce, is "downslope"; i.e., from high to low pressure. In Figure 4-1, if p_1 exceeds p_2, the acceleration will be from right to left. Designating the pressure difference $p_1 - p_2$ as Δp, the magnitude of **P** in Equation (4-1) is given by

$$P = \frac{1}{d} \frac{\Delta p}{L} \tag{4-4}$$

Unlike gravity, which is *always* directed vertically downward, the pressure gradient can act in any direction. However, in the atmosphere, the pressure gradient has a much greater component in the vertical than in the horizontal. Figure 4-2 illustrates a typical pressure pattern in a vertical cross section of the atmosphere. The slope of the isobars in this figure has been greatly exaggerated. For example, typical values for the distances b and c would be 33 m and 300 km, respectively. Thus the slope would be only 33/300,000, or only a 1-km height change in almost 10,000 km. The *horizontal* pressure gradient in this example would be 4 mb/300 km or 0.13×10^{-4} mb/m, while the *vertical* pressure gradient would be 4 mb/33 m, or about 0.12 mb/m. The pressure gradient force in the vertical is thus almost 10,000 times that in the *horizontal*.

The horizontal and vertical components of the acceleration produced by the pressure gradient can be obtained by dividing by the density [Equation (4-3)]. Using the values of the last paragraph and

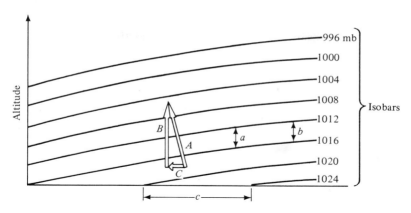

a = perpendicular distance between isobars
b = vertical distance between isobars (exaggerated here)
c = horizontal distance between isobars

A = total or "resultant" pressure gradient
B = pressure gradient (component) in the vertical
C = pressure gradient (component) in the horizontal

FIG. 4-2.   Distribution of isobaric (equal pressure) surfaces in a vertical
cross section.

taking the density as 1.230×10^{-3} g cm^{-3} (Appendix 8), the horizontal
component of the acceleration due to the pressure gradient would be

$$[(0.13 \times 10^{-4}\,\text{mb/m}) \times (10^{+3}\,\text{dynes/cm}^2) \times (10^{-2}\,\frac{\text{m}}{\text{cm}})] \div (1.23 \times 10^{-3}\,\text{g}$$

cm^{-3}) $= 0.11\ \dfrac{\text{cm/sec}}{\text{sec}}$ ($= 0.11$ cm sec^{-2}); the vertical component would

be $[(0.12\ \text{mb/m}) \times (10^3\ \dfrac{\text{dynes/cm}^2}{\text{mb}}) \times (10^{-2}\frac{\text{m}}{\text{cm}})] \div (1.23 \times 10^{-3}\ \text{g}$

cm^{-3}) $= 976$ cm sec^{-2}. Because the vertical portion of the pressure
gradient is so much greater than the horizontal, and since gravity
(which is also quite large) acts *only* in the vertical, it is customary to
treat horizontal forces (and motions) separately from those in the vertical.

In the vertical, there are two relatively large forces: that due to
gravity (about 980 dynes/g) and that due to the vertical pressure
gradient (976 dynes/g in the above example). It is the net difference
between these oppositely directed vertical forces that determines
whether a parcel will accelerate upward or downward, and at what

rate it will do so. Since gravity is essentially constant, the vertical acceleration will depend on the magnitude of the upward component of the pressure gradient force. Since pressure is a measure of the weight per unit area, the change in pressure experienced in, say, a 1-km ascent will depend on the density of the 1-km column of air of unit area. If the layer is very dense, the pressure drop will be greater than if the air is not so dense. At any particular pressure, the density depends on the temperature (Charles' law); that is, warm air is less dense than cold air. (The density also depends slightly on the constituents of the air—dry air is more dense than moist air.) Therefore, the vertical pressure gradient is greater in *cold air* than in *warm air*, and to a lesser extent is greater in *dry* air than in *moist* air.

In summary, there are two rather large "active" forces which drive the atmosphere: pressure gradient and gravity. The latter is directed entirely in the vertical, while the former has a very small component directed in the horizontal. Even though each of the two forces acting in the vertical is much greater than that in the horizontal, this does not mean that vertical motion is much stronger than horizontal motion. It should be remembered that it is the *net* or resultant force that determines accelerations, and the two vertical forces are almost always nearly equal and oppositely directed. Actually, except in small circulation cells such as those in thunderstorms, the vertical velocity of the air is normally only a tenth or a hundredth of the horizontal velocity.

Forces Dependent on Velocity

1. *Coriolis Force.* If one examines the wind pattern on a sea level weather map, it is apparent that the air moves more or less *along* the isobars, rather than directly across them from higher toward lower pressure (Figure 4-3). Evidently a force steers the air flow toward the right of its target (low pressure) in the Northern Hemisphere and to the left in the Southern Hemisphere. This force is the third term on the right in Equation (4-1); it is due to the effect of the earth's rotation and is called the *Coriolis force*, after the French scientist who first explained it mathematically.

The nature of the earth's rotation is illustrated by the turntable of Figure 4-4. The top of the turntable has been given the same sense of rotation (counterclockwise) as the earth's Northern Hemisphere; however, if one looks at the same turntable from below, the sense of rotation is opposite (clockwise). If, now, the turntable is imagined to be as large as a merry-go-round, Figure 4-5 shows what happens if you try to throw a ball from the center of the rotating merry-go-round at a

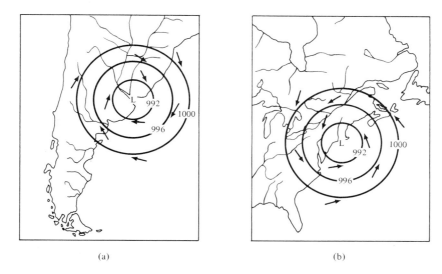

(a)   (b)

FIG. 4-3. Pattern of wind around an area of low pressure. (a) The Southern Hemisphere; (b) The Northern Hemisphere.

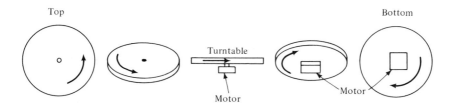

FIG. 4-4. Sense of rotation of a turntable as seen from above and below.

target riding near the edge of the platform. To anyone on the merry-go-round, it will appear as if some force has caused the ball to curve to the right of the intended path in the case of counterclockwise rotation and to the left for clockwise rotation. A spectator not on the merry-go-round would say that the ball moved along a straight line, and that the target moved. The rider on the merry-go-round could account for deflection of the ball from the target by supposing the existence of a deviating force. The same thing happens on the rotating earth: a "straight" line connecting the target (a low-pressure area) and a parcel of air is continuously changing its orientation.

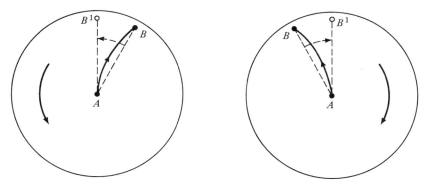

FIG. 4-5.   Effect of rotation on the path of a body. Solid curve *AB* is
the path *relative* to the rotating disk; dashed lines *AB* and
*AB*¹ represent successive orientations of the "target."

On the earth, one may imagine himself on a very large merry-go-
round whose rate and sense of rotation can be varied. This is the situation of the earthbound observer in Figure 4-6. At the North Pole, an
observer is spinning through his vertical axis at a rate of one rotation
per day in a counterclockwise sense. At the South Pole, an observer
is also spinning through his vertical axis at the same rate, but in the
opposite (clockwise) sense. At the equator, the observer is not spinning
around his vertical axis at all, although his "merry-go-round" is turning
about the north-south diameter. At some latitude intermediate between
the pole and the equator, the rotation around the vertical axis is at a
rate between one rotation per day at the pole, and the zero rotation
per day at the equator. It is the rotation around the vertical which
concerns us most, because it is this rotation that affects the horizontal
motion and causes the air to deviate from a direct high-to-low pressure
path.

The magnitude of the Coriolis force can be shown to be:

$$C = 2V\omega \sin \phi \qquad\qquad (4\text{-}5)$$

where ω is the rate of rotation of the earth (one rotation per day or
7.29×10^{-5} rad/sec), V is the speed of an air particle relative to
the earth (i.e., the wind speed), and ϕ is the latitude for which the
calculation is made. (The quantity $\omega \sin \phi$ is the rate of rotation
around the vertical at the latitude ϕ.) The acceleration due to the
Coriolis effect is thus directly proportional to the rotational speed of
the earth, the sine of the latitude, and the speed of the wind. *The
Coriolis force always acts at right angles to the wind.*

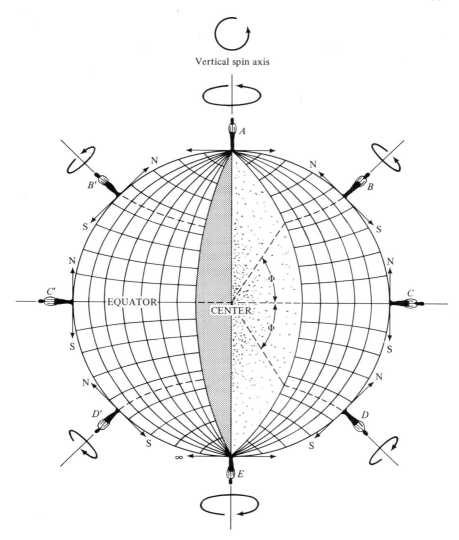

FIG. 4-6. Horizon rotation rate as a function of latitude.

Although the Coriolis force is small, it is significant in horizontal motions of the atmosphere because, first of all, the horizontal pressure gradient force is also relatively feeble, and second, the air traverses great distances. It is very important in large-scale flow, such as that associated with systems that affect the weather over thousands of miles, but it is of much lesser consequence in purely local, small-scale winds. The same, of course, is true of its effect on any body moving freely over the earth's surface. A correction for the Coriolis effect must be applied to long-range artillery, but there is no detectable effect on the path of an ordinary rifle bullet. In the oceans, large-scale currents are appreciably affected, but motions over short distances show no significant deflections.

The Coriolis force becomes effective as soon as a particle has motion. If, for example, a particle had westward velocity in the Northern Hemisphere and there were no forces acting on it other than the Coriolis force, it would follow a path relative to the earth's surface as shown in Figure 4-7. The form of the path would depend only on the speed of the particle and the latitude (the rotational speed of the earth's horizon). When a parcel of air acquires velocity due to a pressure gradient force, the Coriolis force immediately comes into play, constantly deflecting the parcel to the *right* in the Northern Hemisphere, or to the *left* in the Southern Hemisphere. Since we live on the earth and are concerned with the air motion (wind) *relative* to the earth's surface, we can think of this force as being *real* in all respects.

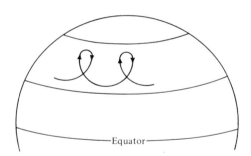

FIG. 4-7. Path of an unaccelerated particle moving relative to the earth's surface.

2. *Friction.* Everyone is familiar with the fact that if a wooden box is given a push along a level floor, it will travel a short distance and then stop. The force that retards the forward motion is friction. It is due to the interlocking of surface irregularities and the adhesion of the molecules of the two surfaces which are in contact.

Although the adhesion is much less, and the space between molecules much greater in a gas than in a solid, there is nevertheless a frictional drag created when velocity differences arise within a gas. This retardation of motion in a fluid is referred to as *viscosity*. When the random, thermal motion of the molecules is responsible for this slowing up, the retardation—sometimes called *molecular viscosity*—is quite low. The decelerating effect of molecular agitation can be explained as follows: If a stream of air is directed along a smooth, solid surface, the air molecules in contact with the surface will have no motion (other than the usual random agitation) because they adhere to the surface. These "surface" molecules will, in turn, retard the flow of those molecules adjacent to them because there is a continuous exchange of the "non-moving" molecules at the surface with those in the next tier. Some of the slow-moving molecules of the second tier mix with those of the third tier, and so on, causing a progressively lesser retardation with distance from the surface.

The molecular viscosity of air is so small that, if it alone were responsible for frictional drag in the atmosphere, the slowing up of the air flow would almost disappear within a meter of the surface. Far more significant is the so-called *eddy viscosity* which, at least in the lower layers of the atmosphere, is about 10,000 times more effective than molecular viscosity in retarding the air motion. As the name implies, it acts through the transfer of momentum between layers of air by *eddies* rather than by molecules.

Eddies are merely parcels of air of various sizes that leave their nomal positions within an otherwise orderly, smooth flow. A wind record marks the passage of these eddies as rapid, irregular fluctuations in both direction and speed. Figure 2-9 shows these fluctuations as recorded by an anemograph. Such fluctuations—deviations from the mean velocity—are referred to as turbulence. In addition to the transport of momentum, turbulent fluctuations greatly accelerate the movement of other quantities in a fluid. The most visible of these are pollutants, such as smoke. As a puff of smoke is carried along by the mean wind, eddies gradually disperse the smoke particles over a larger volume, until the density of the particles is so small that the smoke puff no longer can be seen. Turbulent diffusion also disperses the moisture and heat of the atmosphere.

The intensity of turbulence in the atmosphere depends on several factors. One of the most important is the vertical lapse rate of temperature: the greater the decrease of temperature with height, the greater will be the intensity of turbulence. Turbulence also tends to increase as the wind speed increases, and when the wind blows over rough surfaces. The nature of turbulent flow is illustrated in Figure 4-8. As the effects of the factors listed above increase, the appearance

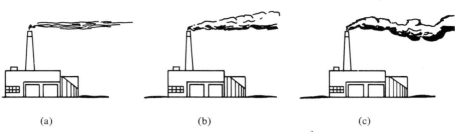

(a)   (b)   (c)

FIG. 4-8.   Effect of turbulence in diffusing smoke. (a) Laminar (non-turbulent) flow; (b) Partially turbulent flow; (c) Well-developed turbulence.

of a smoke plume changes from smooth, laminar flow to increasing eddy motion which disperses the smoke over a larger and larger volume of the atmosphere.

The effect of frictional influences (turbulence) decreases as one ascends higher in the atmosphere. Depending upon the factors mentioned in the previous paragraph, the drag of the surface on the upper flow can extend well above the immediate surface layer. On a day when the atmosphere is well mixed by turbulence, curve *A* in Figure 4-9 shows that the surface can be "felt" as high as 2000 meters. In contrast, Figure 4-9, curve *B*, indicates that when there is little turbulence the surface drag extends upward only 300 meters or less. Under these latter conditions, the speed of the wind at anemometer level (usually 5–10 m above the surface) is only a fraction of the speed in the upper or "free" atmosphere.

Thus, although the frictional force acts in a direction opposite to that of the prevailing wind, its magnitude is a complicated function of altitude above the earth's surface and the degree of turbulent mixing. In the case of large-scale horizontal motions, at altitudes above 1500 meters, meteorologists often omit the frictional force in their computations, assuming that it is much smaller than the pressure gradient force.

3. ***Centrifugal force.***   According to Newton's first law of motion the *velocity* of a body does not change as long as there is no net (unbalanced) force acting on it. But keep in mind that velocity is a vector; i.e., that it has both magnitude (speed) *and* direction. Thus, an object that moves at constant speed along a curved path is changing direction and is, therefore, experiencing an acceleration. This accelera-

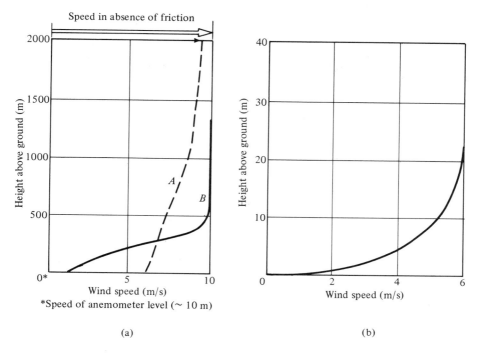

FIG. 4-9. Examples of the variation of wind speed with height in the surface "friction layer." (a) A = strong vertical mixing; B = weak vertical mixing; (b) Variation of wind speed in the first forty meters.

tion is directed at 90° to the path, inward along the turning radius; it is called *centripetal acceleration.*

A centripetal force must be applied to make an object deviate from its natural tendency to move along a straight line. For example, the muscles of a discus thrower apply the centripetal force needed to make the discus move on a circular path until the moment that he releases it; then, it moves along a straight line. We can think of this centripetal force that is making the object follow a curved path as opposed to a *"centrifugal force"* that is apparently pulling the object outward from the desired path.

Air parcels in the atmosphere will experience a centripetal acceleration whenever they travel along paths that are curved relative to the earth's surface. This will occur when the isobars are curved. Although the centripetal acceleration is not specifically named in Equation (4-1), it is really part of the acceleration, **A** on the left-hand side of the equation. When air motion is along a circular or curved path, the acceleration, **A**, can be thought to be composed of two components: one that is directed along (tangent to) the flow and the other at right angles to the flow. The tangential component of the acceleration depends on the rate of change of *speed* of air, while the normal component depends on the rate of change of direction. The magnitude of the normal component of acceleration is given by V^2/r, where V is the wind speed and r is the radius of curvature of the path of the air parcels. For convenience, and because of custom, we shall refer to centrifugal force (which has the same magnitude but acts in the opposite direction) and designate it as **R**:

$$\mathbf{R} = \frac{V^2}{r}\,\mathbf{n} \qquad (4\text{-}6)$$

[(**n**) has been inserted in this equation merely to remind you that the centrifugal force is always directed outward, *normal* to the wind direction.]

In air flow having gentle curvature, such as that in a circulation pattern that covers half a continent or more (Chapter 5), the centrifugal force is generally insignificant compared to other forces. For example, with $r = 1000$ km and $V = 10$ ms^{-1}, $R = (10 \times 10^2)^2/(10^3 \times 10^5) = 10^{-2}$ cm sec^{-2}, which is a tenth the magnitude of the typical pressure gradient force (p. 95). But in the case of a small whirlpool, such as a tornado, in which r may be only 10^4 cm or less, the centrifugal force is very large.

4-3.   The Winds of the Atmosphere

All of the forces which we have described influence the motions that we observe in the atmosphere. However, not all of these forces are equally important under all circumstances. For example, centrifugal and Coriolis forces are relatively unimportant when considering vertical motions. For this and other reasons, we shall consider vertical and horizontal motions separately, beginning with the latter.

Geostrophic Wind

Of the forces we have mentioned, the pressure gradient force is always important for horizontal motion. It is, in fact, the only "driving" force since gravity acts only vertically. On a non-rotating earth, the pressure gradient force would cause the winds to blow directly across

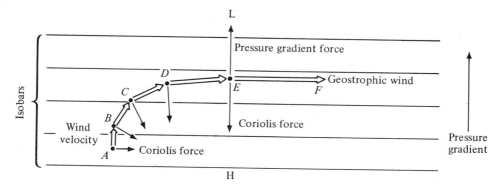

FIG. 4-10. Approach to geostrophic equilibrium by a parcel of air. (N. Hemisphere.)

the isobars, as indicated by the path *A-B* in Figure 4-10. However, because of the earth's rotation, the Coriolis force will deflect the winds to the right in the Northern Hemisphere so that a parcel will take the path *A-C-E-F*, and, in the absence of other forces, the winds will eventually blow parallel to the isobars. This latter condition represents a balance between the Coriolis force and the pressure gradient force. For this particular case, **P** = **C** and substitution of the magnitudes of **P** and **C** from Equations (4-4) and (4-5) yields:

$$\frac{1}{d}\frac{\Delta p}{L} + 2V\omega \sin \phi = 0 \qquad (4\text{-}7)$$

The simple horizontal flow that results from this balance is called the geostrophic wind. It is based on the assumptions that the air parcels are not being accelerated; i.e., that the isobars are straight and parallel (there is no centrifugal force), and that there is no friction. Thus, in Equation (4-1), **A**, **G**, and **F** may be neglected. This kind of "theoretical" air motion is often quite similar to the air flow at altitudes greater than 2 or 3 km above the earth's surface (outside the equatorial latitudes, where the Coriolis force is small).*

Gradient Wind

An examination of the weather map will reveal that there are many instances where the isobars, and therefore the winds, are not straight

---

* See Problem 5 at the end of this chapter.

but curved. In such cases, the winds are not geostrophic, and the effect of the centrifugal force induced by the curved flow must be accounted for. Adding V^2/r, a balance of the three forces would be expressed by

$$\frac{V^2}{r} + \frac{1}{d}\frac{\Delta p}{L} + 2V\omega \sin \phi = 0 \qquad (4\text{-}8)$$

The flow that results from the balance of these three forces is called the *gradient wind*. It is based upon the assumption that the acceleration of the horizontal wind is due only to the curvature of its motion and that there is no friction. Figure 4-11 illustrates the balance of forces and the nature of the resulting flow in the Northern Hemisphere. Where the flow is cyclonic (around the low-pressure center), the wind blows in a curved fashion along the isobars, but the centrifugal force opposes the pressure force. Thus, for the same pressure gradient *the winds in cyclonic gradient flow are weaker than geostrophic.* On the other hand, where the flow is anticyclonic (around the high-pressure center), the centrifugal force aids the pressure force and *the winds in anticyclonic gradient flow are stronger than geostrophic.*

The gradient wind is, of course, a closer approximation to the real flow than is the geostrophic wind. Note also that incorporated into the explanation of gradient wind flow is the requirement that winds in the Northern Hemisphere blow counterclockwise around a low-pressure area and clockwise around a high-pressure area.

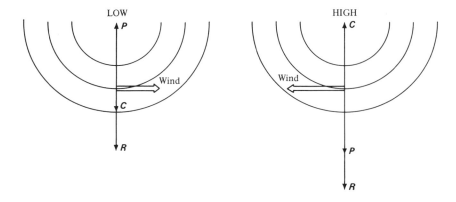

FIG. 4-11.   The balance of forces for the gradient wind (N. Hemisphere). (**P** = pressure gradient force; **C** = Coriolis force; **R** = centrifugal force.)

"Friction Layer" Wind

Near the surface of the earth, the effect of friction cannot be neglected; it is always opposed to the direction of the air motion, and therefore tends to decrease the wind speed. Figure 4-12 illustrates a balance of the four forces (**P**, **C**, **R** and **F**), and the resulting wind.

The effect of adding friction as a fourth force can be seen by considering Figure 4-12. In the case of cyclonic flow (around low pressure), the effect of friction is to reduce the wind speed below that which would exist in gradient flow (Figure 4-11). Thus, the Coriolis force and the centrifugal force will also be decreased, so the relatively greater pressure force will cause the wind to turn inward, across the isobars, toward low pressure. Where the flow is anticyclonic (around high pressure), friction will also reduce the wind speed below the gradient value. Again, the Coriolis and centrifugal forces are thereby reduced, so that the pressure force produces a cross-isobar flow toward lower pressure. Note, moreover, that the centrifugal force does not act in the same direction as the pressure force, but (since the centrifugal force is always directed at right angles to the *wind* direction) at an angle equal to the deviation of the wind from the direction of the isobars.

An inspection of a surface weather map (Figure 8-3) will show that this "friction layer" wind closely describes the wind actually

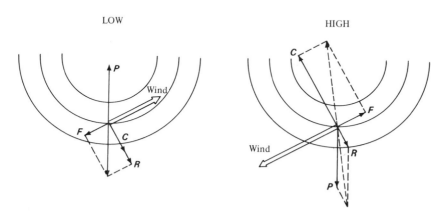

FIG. 4-12. The balance of forces for the "friction layer" wind (N. Hemisphere). (**P** = pressure gradient force; **C** = Coriolis force; **R** = centrifugal force; **F** = frictional force.)

observed in nature near the earth's surface. The amount of deviation of the wind from the isobars depends largely upon the roughness of the friction-causing surface. Over a smooth water or land surface, the deviation is generally 15 degrees or less, while over rough land surfaces, the deviation may amount to as much as 45 degrees or more. The variation of both the direction and speed of the wind within the friction layer is illustrated in Figure 4-13. Note that as the frictional force decreases with height, the speed, and therefore the Coriolis force, increase and the cross-isobar angle, α, decreases.

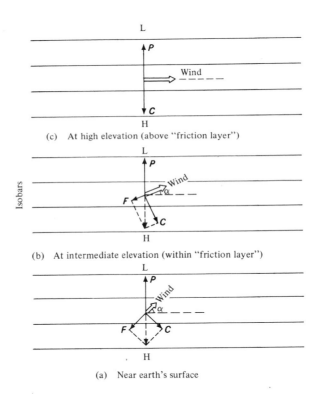

FIG. 4-13.   Wind velocity variation in the "friction layer" (Northern Hemisphere, constant pressure gradient. α is the angle between the wind direction and the isobars.)

Vertical Wind

All of the winds we have examined in the previous sections (the geostrophic wind, gradient wind, and friction layer wind) are restricted to motion parallel to the earth's surface. However, air also moves vertically upward and downward. The driving forces which affect this motion are the pressure gradient in the vertical and gravity. Since the pressure gradient force is always directed upward and the force of gravity downward (Figure 4-14), a balance between the two forces will result in no vertical acceleration of air. However, if the pressure gradient force is greater than the force of gravity, as illustrated in Figure 4-14, the air will be accelerated upward. Similarly, if the gravitational force is larger than the pressure gradient force, the air will be accelerated downward. Of the forces that arise only after there is motion, only friction is significant, and even it has a smaller effect than it has in horizontal motion. Thus, it is the vertical pressure gradient and its variation with height that largely determines the direction and strength of motion in the vertical.

As we have already seen, the pressure gradient force is greatly affected by the temperature of the air. Because warm air is lighter than cold air it will tend to rise, while cold air being heavier will tend to sink. In terms of the balance of the pressure gradient and gravitational forces which we have just discussed, the air will be "lighter" when the vertical pressure gradient force exceeds the force of gravity, and will be "heavier" when the force of gravity exceeds the vertical pressure gradient force. In other words, the vertical forces in the equation of motion [Equation (4-1)] represent a way of describing the buoyancy of the air and the resulting rising or sinking of the air parcels. We shall examine these relationships in more detail in the next section.

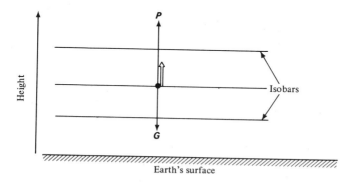

FIG. 4-14.   Forces which produce vertical motion. If the pressure gradient force (**P**) is greater than the force of gravity (**G**), as illustrated, the air parcel will be forced upward (double-shafted arrow).

4-4. Temperature and Air Motions

Thermal Circulation

At the beginning of Chapter 3, the atmosphere was described as a gigantic heat engine. It was stated that the atmosphere's kinetic energy results from temperature differences. In the preceding pages, we have discussed air motions in terms of the pressure gradient force. We must now relate the pressure gradient force with the distribution of temperature in the atmosphere.

Consider the schematic vertical cross sections of the atmosphere shown in Figure 4-15. In (a), the temperature is assumed not to vary in the horizontal at any level, so the density at any given height will be equal everywhere. Thus, the equal-pressure (*isobaric*) surfaces and the equal-density surfaces will be straight and horizontal. But if the air to the south is warmer than that in the north, as in (b), the density at any level will increase from south to north. In this case, the less dense, warm air in the south will rise, while the denser, cold air in the north will sink.

The pressure surfaces near the surface will slope upward toward the colder, denser air of the north. However, because the pressure decreases more rapidly with height in the cold air than it does in the warm (as illustrated in Table 4-1), the north-south slope of the isobars

TABLE 4-1

Altitude Change (Meters) for a One-Millibar Pressure Drop as a Function of Temperature

Pressure (mb)	Temperature, °C				
	−40°	−20°	0°	20°	40°
1000	6.7	7.4	8.0	8.6	9.3
500	13.4	14.7	16.0	17.3	18.6
100	67.2	73.6	80.0	86.4	92.8

will decrease with height. This means that a level will finally be reached where the slope will be reversed. Therefore, although the horizontal flow will always be from high to low pressure, near the surface this means north-to-south (cold to warm) motion, while aloft it will be the opposite, south-to-north direction (warm to cold).

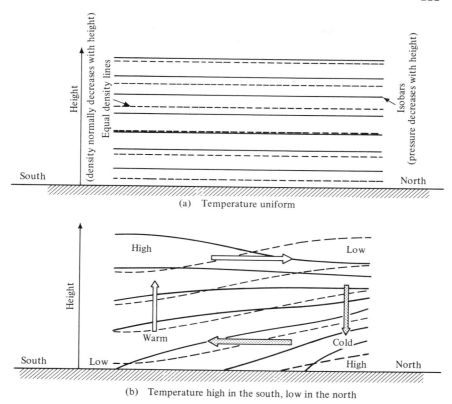

(a) Temperature uniform

(b) Temperature high in the south, low in the north

FIG. 4-15. The principle of thermal circulation.

This "closed circuit" formed by the moving parcels of air is called a *thermal circulation*. It is, in fact, very similar to the motion pattern of water in a pan that has been heated at one edge. Note that this circulation, induced by a horizontal temperature gradient, leads to both horizontal and vertical motions. As we have mentioned earlier, the vertical motion is particularly significant in the production and dissipation of clouds and precipitation. For this reason, it is important that we examine some of the processes associated with the vertical displacement of air.

Adiabatic Processes

Air displaced vertically experiences especially rapid pressure changes. For example, if a parcel of air is forced to descend from an elevation of one mile, the pressure exerted on it will have increased by about 30 per cent by the time it reaches the surface. In response to such a pressure change, the volume and/or the temperature must also change, according to the equation of state [$p = RTd = RT(M/V)$, p. 26]. An indication of what actually occurs can be obtained from our experience in letting air escape from a tire: As the air expands and the pressure drops rapidly, the air cools. This cooling is due to the fact that some of the air's heat energy is expended in doing work of expansion. Conversely, if we pump up the tire rapidly, the air warms because the work we have done in compressing the gas is converted to heat. If we neglect the slow dispersion of heat by conduction through the walls of the tire, we could compute the temperature change by equating the work done in expansion or compression to the change in heat content of the air.

The same process takes place when the pressure of an air parcel is rapidly changed by its descent or ascent in the atmosphere. Assuming that the heat loss or gain through conduction, radiation, and mixing with the surroundings is at a slow enough rate during a vertical displacement (rapid pressure change), the temperature changes can be ascribed primarily to volume changes.

The ideal or theoretical process during which there is absolutely no heat exchange between a gas and its environment is said to be *adiabatic*. However, since air usually contains water, and phase changes may occur that involve the exchange of latent heat, it is necessary to distinguish between two different adiabatic processes: (1) A *dry adiabatic* process is one during which there are no phase changes of moisture; i.e., no condensation, evaporation, fusion, or sublimation. (2) A *moist* or *wet adiabatic* process is one during which phase changes *do* occur, so that account must be taken of the exchange of latent heat.

Since the horizontal pressure gradient is small, and air parcels cross the isobars at a small angle, compression or expansion of air moving horizontally is very slow and may generally be neglected. For this reason, we are concerned only with volume changes of air parcels as they move up or down. Considering first the dry adiabatic process, a displacement of an air parcel in the vertical results in a temperature change of about 1°C for every 100 meters of elevation (or approximately 5½°F/1000 ft). A parcel of air moving upward from near sea level (pressure ~ 1013 mb) in a dry adiabatic process to an altitude of 7000 m (pressure ~ 410 mb) would almost double its volume and its temperature would drop almost 70°C. If the same parcel of air were to be returned dry adiabatically to its original level, its temperature and volume would assume their initial values. Figure

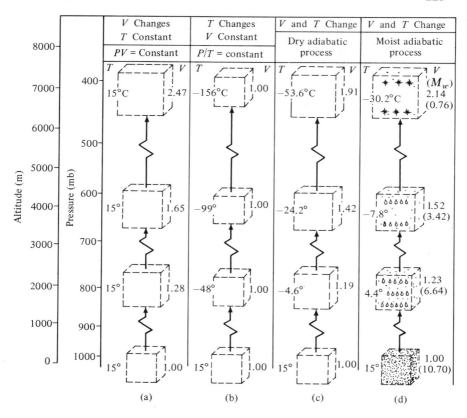

FIG. 4-16. Change in temperature and volume of an air parcel displaced vertically. Only (c) and (d) assume that there is no
heat exchange with the environment. (T = temperature;
V = volume relative to that at sea level; P = pressure;
M_w = saturation mixing ratio in grams of water vapor per
kilogram of dry air.)

4-16, column (c), illustrates the changes experienced by a parcel of air
displaced vertically in the atmosphere in a dry adiabatic process; for
purposes of comparison, columns (a) and (b) show constant-temperature
and constant-volume processes, respectively.

During a moist adiabatic process, the changes of phase of the
moisture contained in a parcel of air undergoing a rapid pressure
change will cause conversion of latent heat to sensible heat, and vice
versa. Specifically, when condensation is occurring, the latent heat
which is released raises the temperature of the air parcel; and when

evaporation is occurring, the latent heat required for the evaporation process cools the air. In column (d) of Figure 4-16, if the air parcel at sea level were saturated, each kilogram of dry air would contain about 10.7 g of water vapor. Then, any lifting of the parcel would cause expansion and cooling, and the excess moisture would have to condense. Thus, when the parcel reached an altitude of 2000 m (pressure ~ 800 mb), the air would still be saturated with water vapor, but the amount of moisture in vapor form would be less than two-thirds the original value.

During this process, more than 4 g of water in each kilogram of air would have condensed. Since the condensation of each gram of water vapor releases about 600 cal of latent heat, the total amount of heat provided to the parcel would be $4 \times 600 = 2400$ cal. Consequently, the temperature of the parcel would be considerably (9°C) warmer than it would have been had the process been dry adiabatic. Further ascent of the parcel would cause more condensation, but because the rate at which condensation proceeds is less when the temperature is low than when it is high, the rate of temperature change during a moist adiabatic process is not constant (as it is in a dry adiabatic process). In Figure 4-16, column (d),[*] this can be seen by comparing the temperature change between 2000 m and 4000 m (about 6°C/km) with that between 4000 m and 7000 m (about 7.3°C/km). Table 4-2 gives values of the moist adiabatic rate at various temperatures and pressures. For practical purposes, at moderate temperatures in the lower layers of the atmosphere, the moist adiabatic rate may be considered to be about 0.5°C per 100 m (or about 3.0 F/1000 ft).

If the "moist" air parcel illustrated in column (d) of Figure 4-16 were to descend, it would warm at the same rate that it cooled, but *only* if all of the moisture remained with the parcel. In practice, however, this is not generally the case. After condensation occurs, many of the liquid or solid water particles leave the parcel of air. Thus, if the parcel descends, much of the original moisture will have been lost to the parcel, so that there will be less evaporation and, hence, a more rapid warming than occurs in a true moist adiabatic process. Whenever some of the condensation products leave the original air parcel, the process is an "irreversible" one and is said to be *pseudo-adiabatic*.

---

[*] The temperature change computed in Figure 4-16 assumes that only condensation occurs in the air. The temperature at which ice particles form in clouds is not fixed. But at the very low temperatures at which they do form, the error introduced by neglect of the heat of fusion is small.

TABLE 4-2

The Moist Adiabatic Temperature Change with Height (°C/100 m)

Pressure (mb)	Temperature °C					
	−60°	−40°	−20°	0°	20°	40°
1000	0.97	0.95	0.86	0.66	0.44	0.32
800	0.97	0.94	0.83	0.61	0.40	0.29
600	0.97	0.93	0.79	0.56	0.36	0.27
400	0.97	0.91	0.73	0.48	0.31	0.24
200	0.96	0.86	0.60	0.36	0.25	0.20
100	0.94	0.77	0.46	0.27	0.21	0.16

Condensation (and the formation of clouds) in the atmosphere is produced principally through the cooling and expansion of air as it ascends into regions of lower pressure. On the other hand, the dissipation of clouds is usually a sign of descending air. The overall pattern of vertical motion which produces a cumulus cloud, such as is illustrated in Figure 4-17, is upward motion below and within the cloud, and downward motion at the edges and outside the cloud. Even air containing little water vapor (low relative humidity) does not require a great deal of vertical lift to create saturation and then condensation. For example, air being lifted from near sea level with an initial temperature of 30°C and relative humidity of only 35 per cent will become saturated at about a 2000-m elevation.

In discussing the characteristics of the troposphere (Chapter 1), it was mentioned that convection keeps this lowest layer fairly well stirred, in contrast to the stratosphere in which there is not very much mixing. Yet, on the average, the temperature in the troposphere is not uniform in the vertical, but rather it decreases at the rate of 6½°C/km (about 3.6°F/1000 ft). The reason for this should now be clear: The air parcels that move up and down in convective currents experience temperature changes that are between the dry and moist adiabatic rates. Thus, the closer the vertical temperature distribution is to being equal to the adiabatic rate, the more thoroughly mixed is the layer.

Vertical Stability

We have already seen how the pressure gradient and gravitational forces are related to the buoyancy and consequent vertical motions of atmospheric parcels. The buoyancy forces can also be described

1356 MST

1406 MST

1416 MST

FIG. 4-17. Stages of development of a cumulonimbus over Arizona. (Courtesy of L. Battan, University of Arizona.)

116

in terms of the "stability" of atmosphere, which is often a more convenient concept with which to deal. A *stable* atmosphere is one in which the buoyancy forces oppose the vertical displacement of air parcels from their original levels; an *unstable* condition exists when buoyancy forces aid the vertical displacement of air parcels; a *neutral* state exists when vertical displacement is neither opposed nor aided by buoyancy forces.

The buoyancy of a parcel of air will depend upon its density relative to the density of the environment at the same level. If a parcel is "heavier" than its environment, it will tend to sink; if it is "lighter," it will tend to rise; if its density is the same as that of the surrounding air, there will be no "Archimedean force" tending to make it either rise or fall.

Since we do not normally measure density directly in the atmosphere, it is more convenient to discuss stability in terms of a quantity which is measured. From Charles' law (p. 25), we know that at any fixed pressure the density is inversely proportional to the temperature. Therefore, at any given pressure level, we can substitute a temperature statement for density in the discussion on buoyancy in the previous paragraph: A parcel of air that is warmer than its surroundings will tend to be forced upward; one that is colder than its surroundings, downward; and one at the same temperature as its surroundings will not experience a tendency in either direction.

We have seen from the discussion of adiabatic processes that the temperature of an air parcel changes at a *fixed* rate when it is displaced vertically; i.e., 1°C/100 m for the dry adiabatic process, and roughly half that rate for the moist adiabatic process. Evidently, then, whether a vertically displaced parcel of air is warmer or colder than the surrounding air at any point along its path will depend upon the vertical distribution of temperature in the surrounding air. This observed vertical temperature distribution in the atmosphere is denoted by the rate at which the temperature decreases with height, and is called the *lapse rate*. For example, the *average* lapse rate in the lower 10 km of the atmosphere is about 6.5°C/1000 m (about 3.6°F/1000 ft).

An atmospheric layer in which the lapse rate is less than the adiabatic is stable. This is demonstrated by Figure 4-18. Consider, for example, a parcel initially at 400 m in the left-hand column, where the environmental lapse rate is 0.7°/100 m. If the parcel is displaced upward from the 400-m level, it will cool at the dry adiabatic rate (1.0°C/100 m) and will always be colder than its environment at the same level. Accordingly, buoyancy will tend to force the parcel back downward again. Or, if the same parcel is displaced downward from

400 m, it will always be warmer than its environment, so that buoyancy will tend to force it back upward. Thus, we see that the environment has a *stable* lapse rate because vertical motions are suppressed.

In the *neutral case* (middle column of Figure 4-18), the lapse rate of the environment is exactly equal to the dry adiabatic rate, and therefore the temperature of the parcel at any new level is the same as the

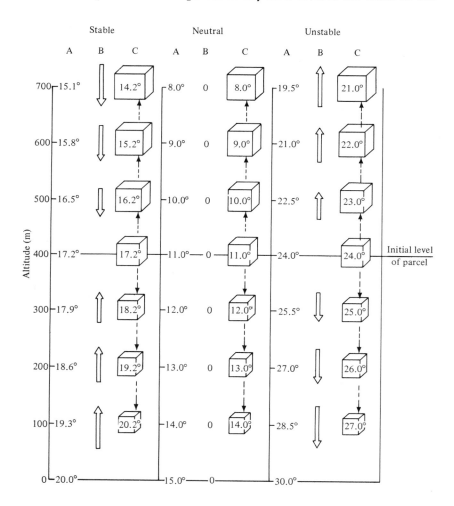

FIG. 4-18. Stability as a function of lapse rate for dry adiabatic processes. (A = environmental air temperature, °C; B = buoyant force; C = displaced air-parcel temperature, °C.)

surrounding air. Thus, it will have no tendency to move either up or down. In the *unstable* case (right-hand column), a parcel displaced upward will be warmer than the environment at each level and, if displaced downward, will be colder than the environment. Accordingly, the lapse rate in the environment is such that vertical motions are encouraged.

A criterion, then, for stability in the case of *dry adiabatic* processes may be expressed:

Lapse rate in environment	*Nature of vertical stability*
Greater than 1°C/100 m	Unstable
Equal to 1°C/100 m	Neutral
Less than 1°C/100 m	Stable

Of course, ascending air currents very quickly become saturated, after which the air cools at the *moist adiabatic rate.* The lapse rate criterion would then be based on the moist, rather than on the dry, adiabatic rate. For example, if the moist adiabatic rate were assumed to be 0.5°C/100 m, and the air parcel in the stable situation shown in Figure 4-18 were saturated, the sounding would be *unstable,* since a lifted parcel would always be warmer than the environment. Thus, condensation of water with its release of latent heat is an important factor in inducing vertical motions in the atmosphere. Further, this means that a significant portion of the heat energy that drives the atmosphere is created through the condensation of moisture.

Changes in vertical stability. Lapse rates in the atmosphere vary considerably both in space and time. An increase of the lapse rate in a layer of the atmosphere results from the warming of the lower part of the layer and/or cooling of the upper part. Conversely, a decrease of the lapse rate is produced by cooling of the lower portion and/or warming aloft. Some of the possible processes which produce different rates of vertical temperature distribution, and thus differences in atmospheric stability, are:

1. *Differential advection.* If the air aloft is being replaced by warmer air brought in by the winds, while in the lower portions cooler air is being brought in, the stability of the air will increase. Conversely, the advection of cooler air at higher, and warmer air at lower levels will decrease the stability of the layer, and an initially stable layer may even become unstable.

2. *Surface heating or cooling.* This can occur in either of two ways: (a) When winds blow over a surface that is either colder or warmer than the invading air. For example, air moving from a cold ocean over a warm continent may cause enough instability to set off showers. In the winter, warm air from the Gulf of Mexico flowing northward over the central and eastern United States sometimes is cooled sufficiently by the cold land so that areas of widespread fog and stratus clouds are produced. (b) When air is over a surface that is losing or gaining heat through radiation. Here, the cooling of the air near the surface on a clear, calm night frequently leads to the creation of a temperature *inversion;* i.e., a lapse rate in which the temperature increases with height (Figure 4-19).

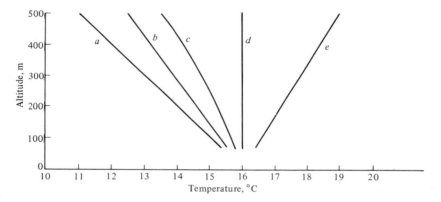

FIG. 4-19. Examples of lapse rates. (a) Dry adiabatic (1°C decrease per 100 m); (b) Normal in troposphere (0.65°C decrease per 100 m); (c) Typical moist adiabatic (about 0.5°C decrease per 100 m); (d) Isothermal (no charge with height); (e) Inversion (increase with height).

3. *Radiative cooling aloft.* Clouds are quite effective blankets, inhibiting the cooling of the earth's surface by radiation to space, and thus retaining the heat in the air below them. However, radiative cooling may take place at the tops of the clouds, and thus produce decreasing stability within the cloud layer itself. Occasionally this process may result in sufficient instability and cloud growth to cause nocturnal showers.

4. *Vertical displacement of atmospheric layers.* When an entire layer of air sinks (goes through a *subsidence* process), the difference in

the per cent compression between the bottom and top of the layer leads to a greater warming of the upper than of the lower portion of the layer. This means that the layer becomes more stable. If the subsidence process is sufficiently prolonged, a *subsidence inversion* may be created. Conversely, if a layer of air is lifted dry adiabatically, its stability decreases. However, if part of the layer becomes saturated during the ascent, the situation changes. If the upper portion becomes saturated before the lower, the upper zone will cool at a lesser rate than the lower, and the stability will be increased. But if the lower portion becomes saturated earlier, then the upper part will cool faster than the lower, and the stability will be decreased. Thus, instability is likely to occur in a lifted layer of air if the bottom of the layer is relatively moist while the top is relatively dry.

4-5. Relationship between Horizontal and Vertical Motions

Although we have treated vertical and horizontal motions separately, the two components are, of course, always related, as we pointed out in the discussion of the thermal circulation. The upward and downward convective currents induced by instability, for example, must always be connected by horizontal flow between them. In other words, regardless of the *size* of the circulatory "cell," there must be continuity of flow; i.e., a closed circuit must be formed to prevent the creation of "holes" in the atmosphere. Such closed circuits may have horizontal dimensions of only a few miles, as in the case of small cumulus clouds, or thousands of miles, as in the case of the large weather-map "highs" and "lows."

On a large scale, the pattern of upward and downward motion can be related to the horizontal air currents that circulate around the great cyclonic and anticyclonic whirls of the atmosphere as follows:

The air currents that spiral inward at low levels of cyclones converge; i.e., the winds blow inward toward the center (Figure 4-12). Since the horizontal area occupied by a volume of air must therefore decrease with time, the vertical depth must increase. This is illustrated by the upper half of Figure 4-20. Imagine a column of air, having the boundaries shown in the figure, with streams of air spiraling inward toward the center. This inward flow results in a shrinking of the horizontal cross-sectional area with time (*convergence*). Since the total mass of air in the column must remain constant, it follows that its height must increase. Air must therefore move *upward* within the

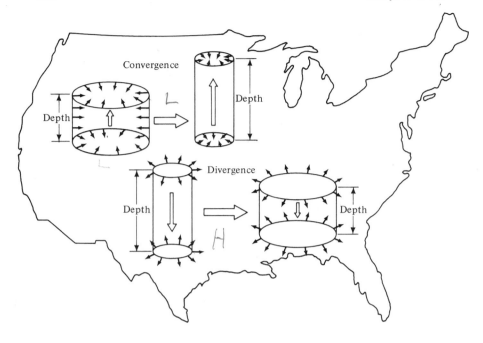

FIG. 4-20. Convergence and divergence of a disk of air.

column. In the columns labeled "divergence," the outward flow of air illustrates what takes place in an anticyclone. Here, the outward flow results in an expansion of the column's horizontal cross-sectional area (*divergence*), and vertical shrinking must take place in order that the total mass be kept constant. Air must therefore move *downward* in the column.

At levels above 6 or 7 km, compensating horizontal convergence occurs over surface anticyclones, while divergence occurs over surface cyclones. Thus, a pattern of motion such as that illustrated in Figure 4-21 takes place. It should be noted, however, that the vertical scale in this drawing is greatly exaggerated. The diameter of a typical cyclone or anticyclone is greater than 1000 km, so that the slopes of the upward and downward flows are not nearly as steep as they appear in Figure 4-21. As a matter of fact, not only do the air parcels move more gradually upward and downward, but the vertical velocities produced by these large-scale patterns of convergence and divergence are usually not more than a few centimeters per second (1 mile per day). However, they are sufficient to set the weather "stage" over large areas. In the absence of other influences, the weather over areas dominated by cyclones tends to feature widespread cloudiness and precipitation, while that over anticyclonic areas is usually clear.

This relationship between the vertical and horizontal motions of the atmosphere and the "weather" holds equally well for the whole hierarchy of atmospheric whirls of different sizes. We shall examine the nature and magnitude of all the significant circulation patterns in the next two chapters.

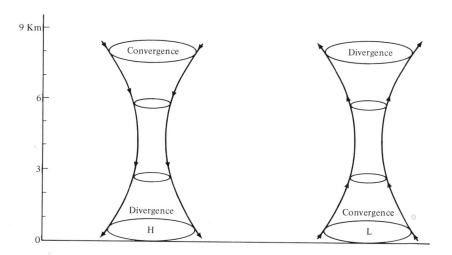

FIG. 4-21. Large-scale patterns of divergence and convergence.

PROBLEMS

Problems marked with an asterisk (°) are the most challenging.

1. If the earth were not rotating, what would be the direction of the horizontal wind with respect to the isobars? What force would keep the air from increasing its speed ad infinitum?

2. Explain the behavior of a Foucalt pendulum. At what rate does it rotate? (Review the discussion of the Coriolis force.)

°3. The gravitational force, according to Newton's law, is given by KM_1M_2/r^2, where K is the universal gravitational constant, M_1 and M_2 are the masses of the attracting bodies, and r is the distance between the masses. Explain the statement on p. 93 that this force varies with both latitude and altitude. "Apparent gravity" combines the effects of gravitational attraction and centrifugal force. How does "apparent gravity" vary with latitude? (Hint: Draw a circle to represent the earth and at each of several latitudes between the equator and pole show an arrow to represent the direction and strength of "true" gravity and a second arrow to represent the direction and strength of the centrifugal force due to the earth's rotation. Combine the two arrows at each latitude into one.)

4. For a given wind speed, how much stronger (or weaker) would the Coriolis force be at a latitude of 45° than at 30°? Than at the pole? What is the magnitude of the Coriolis force acting on any wind at the equator?

5. Can there be a geostrophic wind at the equator? Explain your answer.

6. Since the Coriolis force in the Northern Hemisphere always tends to deflect a moving air parcel to the right, why do the winds blow cyclonically (counterclockwise) around a low-pressure area?

7. Which of the forces in the equation of motion, Equation (4-1), are influenced by the effect of friction? Which forces are always perpendicular to the wind direction? Which are always perpendicular to the isobars?

8. A balloon which is made of a non-expansible plastic is filled with helium and released into the atmosphere. If it stops rising and floats near the 2000-m level, what can you say about the average density of the balloon and its contents relative to that of the air at the same level? How would you decide how much helium to put into the balloon before releasing it?

9. Before light gases such as helium were available in quantity, balloonists inflated their balloons with hot air. What does this show about the effect of temperature on density, and thus also the buoyancy? Why does smoke normally rise in the atmosphere?

°10. From the differences in temperature change between the dry and moist adiabatic lapse rate (Figure 4-16) over various altitude increments, compute the approximate specific heat of moist air. How should the specific heat change with increasing mixing ratio? (Note: The latent heat of vaporization is about 615 cal/g at −30°C; 601 cal/g at −8°C; 595 cal/g at 4°C; and 589 cal/g at 15°C. The specific heat of dry air is about 0.240 cal/g/°C while that for water vapor is 0.45 cal/g/°C.)

11. Is the *average* lapse rate in the atmosphere (6.5°C/km) stable with respect to the *dry* adiabatic rate (10°C/km)? Is the average lapse rate stable with respect to a *moist* adiabatic lapse rate of 5°C/km?

Since the air almost always contains some moisture, what does this suggest about the average stability of the troposphere?

12. Under what conditions of moisture distribution within a layer of air will convergence at the surface and divergence aloft tend to decrease the stability of the layer? Will convergence aloft and divergence at the surface tend to increase or decrease the stability of a layer?

13. Plot a graph having as the abscissa, temperature over the range of $+30°C$ to $-55°C$; and ordinate, height over the range 0 to 16 km. On the right-hand vertical scale, indicate the standard pressure in the vertical, as determined from Appendix 8 and Figure 1-1. Plot the following three temperature-pressure soundings measured in three different air masses by connecting consecutive points in each sounding with straight line segments. Label each curve. Draw several straight, sloping lines on the chart to illustrate the rate at which temperature changes with height during a dry adiabatic process, one starting at sea level and 30°C, another at 0°C and sea level, and a third at $-30°C$ and sea level.

   (a) Identify layers in the three soundings that exemplify absolutely stable, absolutely unstable, and neutral stratifications.

   (b) Identify all layers containing either an inversion or isothermal lapse rate.

   (c) Where would you say the tropopause is located in the first two soundings?

Air mass:	Tropical	Polar (summer)	Polar (winter)
Pressure (mb)	Temp. (°C)	Temp. (°C)	Temp. (°C)
1000	27	13	−31
950	22	13	−32
900	25	9	−32
850	22	4	−30
800	20	−1	−30
700	13	2	−28
600	6	−8	−31
500	−1	−17	−38
400	−13	−32	
300	−28	−50	
200	−50	−50	
150	−51	−50	

5

Atmospheric Motions: Circulation Patterns

5-1. Scales of Motion

In the last chapter, it was shown that temperature differences produce the basic forces that drive the winds. However, the picture of the air flow is made complex by the earth's rotation, frictional drag and turbulence, mountain obstacles, and perhaps most of all, the incessant changes of the state of water in the air. To simplify the analysis of the enormously complex patterns of "eddies within eddies" that exist in the atmosphere, it is convenient to categorize circulation systems according to size. Almost every size is represented in the atmosphere: everything from the very small whirls that kick up the dust on a road to enormous oscillations that have horizontal dimensions of several thousand kilometers. All of these different sizes—or *scales of motion,* as they are called—are interdependent. For example, an eddy produced by a hill might not occur unless there was a prevailing wind due to a circulation of much larger size.

An instantaneous snapshot of the winds of the entire atmosphere would present an extremely chaotic view of the flow. The complex distribution of forces producing such flow would make prediction an impossible task. To achieve some order, a type of filtering, according to size of flow elements, must be applied. This can be accomplished by a system of averaging. For example, the very small-scale eddies or whirls that cause a windvane to oscillate rapidly or branches of a bush to sway with periods of perhaps only a few seconds can be eliminated by averaging the observed wind velocity over periods of several minutes. If one were to average the wind velocity over an entire day, then wind oscillations having periods of much less than a day would disappear from the record. Meteorological observations are averaged both over time and space to isolate the various sizes of atmospheric motions. The analyst of the weather maps that are published in the newspaper applies an averaging process—a smoothing of isobars that eliminates most irregularities smaller than about 100 km.

Actually, most routine meteorological measurements are made in such a way that very small eddies are eliminated. Most anemometers and thermometers do not react to small, high-frequency changes. Observations are so widely spaced, both in time and area, that most must be considered averages over horizontal distances of tens of kilometers and vertical distances of tens of meters. Even such relatively large circulation phenomena as thunderstorms and tornadoes often fall through the "mesh" of the usual weather-station network.

The scales of atmospheric motions can be classified as shown in Table 5-1. The wind observed at any place can then be thought of as a composite of several different scales of motion, each having somewhat different dominant causes. For example, the macroscale flow patterns are associated with the large features of the earth's surface and distribution of heat: continents and oceans, extensive mountain ranges, latitudinal variations of insolation. The various scales of motion can also be characterized by the magnitude of the vertical motion associated with each. Macroscale motion is mostly in the horizontal: The vertical displacements attributable to the very large circulation features are no more than 1 or 2 cm sec^{-1} (0.02–0.04 mph); even in the great cyclonic storms that regularly affect the middle and high latitudes, average vertical displacements are usually less than 50 cm sec^{-1} (1 mph). In the smaller, more intense mesoscale circulations, the vertical velocities are often more comparable to the horizontal velocities;

TABLE 5-1

Scales of Motion in the Atmosphere

Typical horizontal dimension	*Description*
A few centimeters to a few kilometers (microscale)	Small high-frequency eddies, often referred to as turbulence, that are strongly affected by local conditions of both terrain roughness and temperature. The lifespan of individual eddies is usually less than a few minutes. Very significant as diffusers of pollutants in the air. Coriolis force is not significant.
1–100 km (mesoscale)	Small convective cells that persist for many minutes or hours, such as the land-sea breeze, mountain-valley breeze, tornadoes, thunderstorms. Coriolis force generally not significant.
Hundreds to several thousands of kilometers (macroscale)	The cyclones and anticyclones that are largely responsible for the day-to-day weather changes. Such systems persist for days or even weeks. Coriolis force very significant.
Few thousand to 10,000 km (large macroscale)	Features of the atmospheric circulation that persist for weeks or months. Long waves that exist in this flow move very slowly or not at all across the earth. These play an important role in the characteristics of weather over periods of a month or more. Coriolis force very significant.

for example, in a thunderstorm, the vertical motion is often 10 m sec^{-1} (22 mph) and can reach 30 m sec^{-1} or more. Winds of the microscale are generally much weaker than those of the larger-sized motions, but the vertical motions are very nearly equal to those in the horizontal. However, microscale motions, unlike those of the mesoscale, appear to occur principally in a rather shallow layer adjacent to the earth's surface.

5-2. The Nature of Atmospheric Circulations

At the beginning of Chapter 3, attention was drawn to certain simi-
larities between the heat engine and the atmosphere. This analogy can
be demonstrated in yet another way. Just as the motor of an auto-
mobile functions through the turning of wheels and gears of various
sizes, the total kinetic energy of the atmosphere is partitioned among
circulations of varying dimensions. Furthermore, in the mechanical
engine, the larger the mass of the rotating wheel or gear, the greater
the power required to turn it. In the atmosphere, also, the amount of
energy required to initiate and maintain a circulation pattern is, for
the most part, directly proportional to the mass of air associated with
the circulation.

The wind system which contains the greatest mass of air is called
the *general circulation.* Driven by the energy received from the sun,
it serves to transport the air from the equatorial regions toward the
poles, and to maintain a return flow of cold air from polar to tropical
latitudes. It determines, in large measure, the broad pattern of climates
of the earth. Within this large-scale global flow are embedded the
smaller circulations. These smaller circulations—"perturbations" on the
worldwide flow—are responsible for the transient, short-period variations of atmospheric conditions; i.e., the weather.

In this chapter, we shall examine the structure and causes of various
sizes of atmospheric circulation patterns.

The General Circulation

The general circulation of the atmosphere is the average flow of air
over the entire globe. It is determined by averaging wind observations
over long periods of time—usually twenty years or more. In order to
isolate the seasonal variation of the general circulation induced by the
earth's revolution about the sun, the averaging is sometimes done separately for each season of the year. In any case, this long-period averaging tends to eliminate the smaller circulations, as we explained at
the beginning of this chapter.

If the earth were not rotating and if the surface were homogeneous,
solar heating at the equator would cause the air in that region to rise
and flow toward the poles. As it was transported poleward, not only
would the air become cooler and tend to sink toward the surface, but
the convergence of the meridians of longitude would force the air to

"pile up" before it reached the pole. These effects would induce a return circulation near the earth's surface from polar regions to the equator. In practice, however, this simple thermal circulation pattern between the poles and the equator is greatly modified by the rotation of the earth, and by the nonuniform properties which are characteristic of its surface. Instead of a single circulation cell from equator to pole in each hemisphere, there are three latitudinal circulations, and there are also important longitudinal variations around each hemisphere.

The picture of the general circulation can be simplified somewhat by averaging the observed winds along each latitude, thus eliminating the longitudinal variations. Figure 5-1 is a schematic representation of the results. The horizontal flow at the earth's surface is shown in the center of the diagram; the net meridional circulation, both at the surface and aloft, is depicted around the periphery. The component of the flow along meridians (north/south) has a speed, on the average, of less than a tenth of that along latitude circles (west/east), indicating the importance of the effect of the Coriolis force on this largest of scales.

Within the equatorial region are the *doldrums*, a belt of weak horizontal pressure gradient and consequent light and variable winds. Here, also, is the region of maximum solar heating, so that the surface air rises (as shown by the vertical circulation at the edge of the diagram), and flows both northward and southward toward the poles. This poleward flow at high levels is acted upon by the Coriolis force, turning the wind to the right in the Northern Hemisphere, and to the left in the Southern Hemisphere. Thus, in both hemispheres, the poleward flowing air becomes a west wind and, at an average latitude of about 30°, reaches a maximum speed which may exceed 100 mph. These are the *jet streams*, which are discussed in more detail on p. 137.

At about 30° north and south latitudes, some of the air descends again toward the surface. This is the region of the *horse latitudes* (so-called because Spanish sailing vessels, carrying horses to the New World, were occasionally becalmed in these areas of light winds and many of the animals had to be thrown into the sea because of the lack of food). Since the air is generally descending in these zones, there is little cloudiness or precipitation, and it is here that most of the world's great deserts are found.

Between the doldrums and the horse latitudes are wide belts where a portion of the previously equatorial air, having been cooled by its journey northward and dried out by its descent to the surface, returns again to the tropics. However, it does not flow directly southward, but

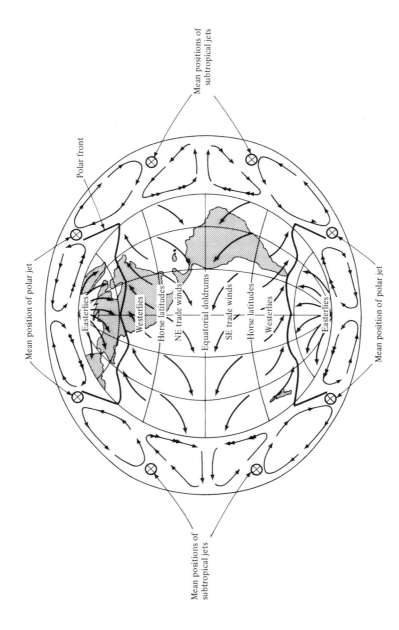

Mean positions of subtropical jets

Polar front

Mean position of polar jet

Easterlies
Westerlies
Horse latitudes
NE trade winds
Equatorial doldrums
SE trade winds
Horse latitudes
Westerlies
Easterlies

Mean position of polar jet

Mean positions of subtropical jets

FIG. 5-1. Schematic representation of the general circulation of the atmosphere. Double-headed arrows in cross section indicate wind component from the east.

is deflected by the Coriolis force, so that the wind moves from the northeast in the Northern Hemisphere and from the southeast in the Southern Hemisphere. These are the remarkably persistent *trade winds*, which obtained their name because of the important role they played in opening up the New World when ships were dependent upon sails.

From the horse latitudes, some of the descending air moves poleward at low levels, but the flow is deflected by the Coriolis force and the winds have a westerly component. These are the *prevailing westerlies*. During the days of sailing ships, they provided the motive power for vessels returning from North America to Europe. As the warm poleward-moving air reaches a latitude which varies from 40° to 60°, it encounters a cold flow from the pole. As a result of this encounter, a boundary is formed between the two masses of air known as the *polar front.*° Here, the warm, light air from the horse latitudes is forced to rise over the cold, dense air from the pole, and a portion of the warm air returns at high levels toward the equator. Although the Coriolis force tends to deflect the returning air toward the east, the return flow already has sufficient westerly momentum to overcome the Coriolis effect, and the flow in this zone remains westerly at all levels.

Poleward of the polar front are the *polar easterlies*. These winds bring the cold arctic and antarctic air from the polar regions toward the polar front, where they are warmed by their equatorward movement (and also in individual storms by mixing with the warmer air on the other side of the front). The air thus rises and returns toward the pole as a westerly flow aloft.

From this description of the general circulation of the atmosphere, we see that there are two primary zones of rising air—in the tropics and in the region of the polar front. As might be expected, it is here that the principal areas of precipitation are found. Complementing these regions are the zones of descending air—in the horse latitudes and near the poles. Here, the precipitation is relatively light. We have already noted that the major deserts of the world are found in the horse latitudes and, while meteorological data near the poles are sparse, it is known that the precipitation in these areas is also small. However, because of the very low rate of evaporation of the ice in polar regions, it remains on the ground for long periods of time.

The preceding discussion represents an average or *mean* description of the atmospheric circulation for the entire year as a function of latitude only. The patterns of mean sea level pressure during January and July shown in Figure 5-2 give some idea of how the circulation

---

° We shall discuss the polar front in more detail in the next chapter.

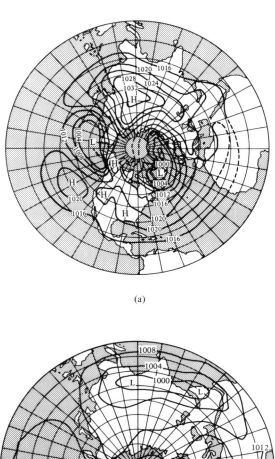

(a)

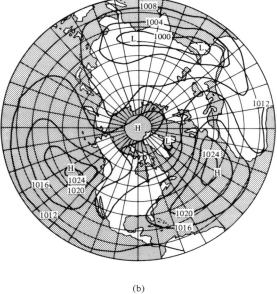

(b)

FIG. 5-2. Normal sea level pressure in the Northern Hemisphere. (a) January; (b) July.

134

varies over the surface of the earth during the year. Note that the subtropical high-pressure belt, associated with the accumulation of air in the horse latitudes, is not continuous around the hemisphere either in winter or summer, but rather is broken up into cells over the Atlantic and Pacific Oceans. These two cells are especially well defined in summer, and are displaced several degrees of latitude farther northward in summer than in winter.

Since the winds blow clockwise around high pressure in the Northern Hemisphere, the *eastern* periphery of each cell is under the influence of relatively cool, dry northerly flow. Thus, the coastal areas of southwestern North America and Europe are favored by generally pleasant, rainless summers. In the interior of these regions (including northern Africa) are the major deserts. The *western* periphery of each high-pressure cell is associated with warm, moist flow from the tropics. Accordingly, locations such as the southeastern United States, as well as Hawaii, the Philippines, and southeastern Asia, are typically warm, with high humidity and frequent summer showers.

Near the latitude of the polar front, where the relatively warm, moist prevailing westerlies meet the cold polar easterlies, are two low-pressure centers. These are well defined in winter, but almost disappear in summer. Because of their location, they are termed the Aleutian low (in the Pacific) and the Icelandic low (in the Atlantic). They represent semipermanent "centers of action," where the major midlatitude storms develop their greatest intensity.

In the interior of the North American and Asian continents, the low temperatures of winter result in increased density of the air at low levels and produce the cold high-pressure systems noted in those areas. However, in summer, temperatures are high, the air is less dense, and warm low-pressure systems prevail.

Figure 5-3 shows the corresponding flow at 500 mb (about 18,000 ft) during the summer and winter in the Northern Hemisphere. At these levels, westerly winds dominate the region poleward of the horse latitudes, but are more intense in winter than in summer, as evidenced by the closer spacing of the winter contours. The geographic center of the flow (lowest contour height) is not generally located at the pole, but is displaced some distance away. In the Northern Hemisphere, the center of the lowest 500-mb height is located over western Greenland, with a secondary center over eastern Siberia. These centers are the upper-level reflections of the Icelandic and Aleutian lows, referred to earlier in this section.

Since the oceans in the Southern Hemisphere cover a significantly larger area than the oceans in the Northern Hemisphere, the tempera-

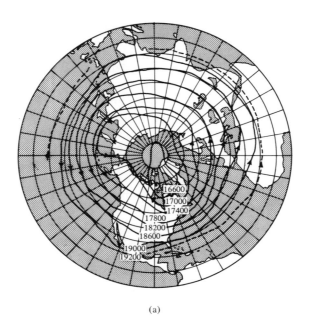

(a)

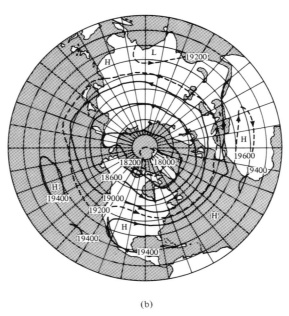

(b)

FIG. 5-3. Normal contours at 500 mb in the Northern Hemisphere. (a) January; (b) July.

ture distribution is much less affected by fluctuations due to continental influences. Thus, the pressure patterns, and therefore the large-scale winds, are much more symmetrical around the pole in the Southern Hemisphere than in the hemisphere north of the equator. Otherwise, the upper-level flow is similar, being consistently from the west throughout the region from about latitude 30° to near the pole.

Although the existence of *jet streams* had been postulated by theory much earlier, their actual existence was not observed until 1946 when high-flying military aircraft encountered unexpected strong head winds, against which they could make but little progress. Figure 5-4 shows the average positions of the jet streams. The maps show the wind speeds associated with these "rivers of air"—narrow bands of high-velocity winds—during the winter and summer seasons in the Northern Hemisphere. While the mean speeds shown on these charts only slightly exceed 100 mph, extreme winds in excess of 250 mph have been recorded in individual instances.

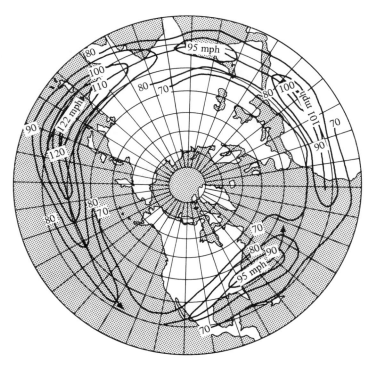

FIG. 5-4. The jet stream in winter. Arrows designate axis or "core" of jet. (After Namias and Clapp.)

The location and intensity of jet streams change from day to day throughout the year. They are associated with zones of strong horizontal temperature gradients, and therefore follow closely the oscillations in position and strength of the polar front. In addition to the circumpolar jet stream of middle latitudes, other jet streams have been observed. A "subtropical jet stream" occurs in lower latitudes at very high elevations (9–13 km), while a "polar jet" exists during the winter season above the arctic regions at elevations exceeding 20 km.

Basis for the general circulation. Our knowledge of the general circulation is based on a sparse observation network. Very few observations have been made over the unpopulated areas of the earth—the oceans, mountains, jungles, arctic, and antarctic—which comprise most of the earth's surface. Nevertheless, sufficient information has been gathered to indicate that the general circulation is considerably different from that which might be expected for a uniform, nonrotating earth.

Clearly, a single thermal cell does not extend from equator to pole. The meridional (south-north) component that would be expected in a simple thermally driven circulation has become mostly "zonal" (along latitude circles) because of the influence of the Coriolis force. There is only one remnant of a clearly defined thermal circulation—that of the tropical cell* between the equator and 30°. There may also be a much weaker and less persistent *polar cell* over polar regions. But certainly between these two zones, and perhaps over the entire earth poleward of 30°, there is no organized thermal circulation; rather, only frequent large eddies (cyclonic disturbances) that intermittently transport heat and momentum between the tropical cell and the polar regions. Note from Figure 5-1 that the mean flow within the middle-latitude cell† shows meridional motion that is actually opposite to what a thermal circulation should be like.

A study of the angular momentum of the atmosphere indicates that air moving poleward from the tropics would have an excessive westerly speed if there were not some mechanism by which its momentum were reduced. This can be seen by recourse to a basic principle of physics. The angular momentum of any body is given by $m\omega r$, where m is the mass of the body (in this case a particle of air), ω is the angular velocity

---

* Sometimes referred to as the *Hadley cell*, after the person who first proposed, in 1735, that such thermal meridional circulations extend between the equator and the poles.

†Sometimes called the *Ferrel cell*, in honor of William Ferrel, an American meteorologist.

of the body, and r is the "spin radius"; i.e., the perpendicular distance of the body from the earth's axis.

Now if there are no torques (forces that cause turning, such as that applied to a pipe by a wrench), the angular momentum of the air parcel will not change. But if a parcel of air moves from a lower to a higher latitude at a constant distance from the earth's surface, its distance from the earth's axis (r) will decrease. Thus, if its angular momentum and its mass (m) remain constant, the angular velocity, ω, must increase. (This is like the skater who makes himself spin faster by pulling his arms in toward his body, thus concentrating his mass near the spin axis.) Since the earth's angular velocity is the same at all latitudes, if the parcel were originally spinning at the same rate as the earth's surface, it will be spinning at a faster rate than the underlying surface when it arrives at higher latitudes. This means that, as an air parcel moves from the equator toward the pole, it will have acquired an additional speed from the west, relative to the earth. As an example, a parcel of air displaced from the equator to latitude 60° would acquire a west-to-east speed of about 230 m/sec (515 mph)! Since such speeds are far greater than those ever observed, some mechanism must be responsible for slowing down the air which is transported from the equator to higher latitudes.

This mechanism is believed to be the cyclonic disturbances, or storms, of middle latitudes. The air in the easterly trade winds, slowed by the friction of the earth's surface, has its westerly angular momentum increased as it moves toward the equator. This acquired angular momentum, transported poleward by the tropical cell, is gradually absorbed by the great cyclonic storms of middle latitudes and, in turn, is dissipated at the earth's surface through friction. Thus, these storms embedded in the westerly flow of middle latitudes dissipate the excess momentum, much like the small turbulent eddies near the earth's surface diffuse a high concentration of smoke in the air.

While an overall analysis of the general circulation can thus be made from our current knowledge, a more detailed explanation of its time and space variations cannot yet be accomplished. This means that, since predictions of the weather for weeks or months in advance necessitate a better understanding of the behavior of this largest scale of motion, accurate day-to-day, long-range weather forecasting cannot be achieved at present. However, modern technological developments such as meteorological satellites, electronic computers, and advancements in meteorological theory all give promise that significant progress can be made toward this goal.

Circulation in the stratosphere and mesosphere. The general circulation described in the previous paragraphs is that of the troposphere. Above the troposphere, our knowledge of the air circulation is even poorer. Weather balloons (radiosondes) often reach altitudes of 30–35 km, but in much smaller numbers than is needed to obtain a clear picture of the wind and temperature patterns of the upper atmosphere; for altitudes greater than 30 km, we must rely on rockets, which are sent aloft at relatively few locations on the globe. Nevertheless, the number of observations has increased markedly during the past decade or so, and we now have at least a fair idea of the circulation within the stratosphere and mesosphere.

As we have pointed out in earlier chapters, the troposphere receives its energy principally at the earth's surface, with convection transporting heat throughout the layer. Convection to levels beyond the tropopause is strongly inhibited by the stability of the stratosphere. (Recall from Chapter 1 that the temperature increases with height in the stratosphere and, from the discussion in Chapter 4, that such a temperature lapse rate tends to oppose vertical motion.) Thus, energy transfer above the tropopause is largely by means of radiation. It follows, then, that the energy distribution in the upper atmosphere is closely related to the radiative properties of its constituents and the intensity of incident radiation.

Since radiation is so important, we should expect isotherms to be aligned even more nearly along latitude circles than they are within the troposphere. The isobars, too, tend to be more nearly along latitude circles and, thus, the winds are largely zonal; i.e., from west to east or east to west.

The average speed of the zonal component of the winds up to 100 km is shown in Figure 5-5. The most striking characteristic of the circulation is the reversal of the wind direction in the stratosphere from easterly in the summer to westerly in the winter. The easterlies reach their peak in late July and early August, the westerlies are at a maximum in late January; the spring reversal occurs in April or May and the fall transition is in late September or October.

The circulation of the lower stratosphere (below about 25 km) is largely dominated by that of the troposphere. The large cyclonic and anticyclonic systems of the troposphere are still in evidence in the lower stratosphere. On the average, the tropopause slopes downward with increasing latitude from about 17 km over the tropics, where its temperature is about −80°C, to 10 km or less over the polar regions, where its temperature is about −50°C. In the lower stratosphere, the

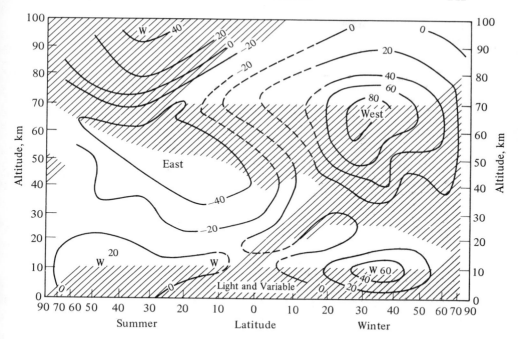

FIG. 5-5. Average zonal (west to east) component of the wind during summer and winter and layers in which temperature decreases poleward (shading). (Northern Hemisphere, speed in ms^{-1}; after Batten and Teweles.)

temperature actually increases poleward at all heights, which is just the reverse of the situation in the troposphere. During the summer, when the daily insolation received in polar regions is almost equal to that of tropical regions, this increase of temperature with latitude is found throughout the hemisphere; but in winter, when the insolation received in polar regions is sharply reduced, the temperature increases only to a latitude of about 60° and then decreases poleward. (This is illustrated by the shaded and unshaded areas of Figure 5-5.)

Because of this reversal in the horizontal temperature gradient, the equator-to-pole pressure gradient that dominates poleward of 10°–15° latitude in the upper troposphere weakens considerably in the lower stratosphere. At 25 km, the circulation is very weak in all seasons— from the west in winter and from the east in summer.

Although the tropopause is nominally the limit of convective mixing in the troposphere, there is evidence that there is some exchange of air across this boundary. Although there is no clear indication as yet just how much transport there is across the tropopause, studies have shown that stratospheric air is frequently injected into the storm systems of middle latitudes in thin filaments.

A curious, approximately biennial, cycle of winds was discovered at the end of the International Geophysical Year (July, 1957–Dec., 1958) in the lower stratosphere over equatorial regions (Figure 5-6). This periodic reversal of the winds begins just above the tropopause, is most pronounced near 24 km, and extends to at least 30 km and probably 40 km or more. Its amplitude is greatest at the equator, decreasing poleward up to about 25° latitude; a change in phase (time of the easterlies) occurs near 25° latitude and the amplitude of the oscillation is a fifth or less what it is over the equator. At low latitudes, the wind reversals first appear at very high elevations and gradually propagate their way downward, at a rate of about 1 km per month. There is as yet no adequate explanation of the phenomenon. Biennial cycles of certain climatic elements have been found which may be associated with this mysterious circulation change.

The upper stratosphere (25–50 km) is very stable, having a temperature increase with height in tropical regions of about 2.5°C/km. Its heat is derived largely through absorption by ozone of the sun's ultraviolet energy, as we explained in Chapter 1. The warmest region, therefore, is closely associated with the angle of the sun's rays. As can be seen from Figure 5-5, the temperature in this layer decreases poleward during the Northern Hemisphere winter and reverses itself in the summer. Since pressure decreases most rapidly with height in cold air (Chapter 4), low pressure appears over the winter pole; this low center intensifies up to the stratopause. In summer, with relatively high temperature over the pole, there is high pressure over the pole. In addition to this seasonal (monsoon) wind circulation there is a diurnal variation associated with the daily changes in solar intensity.

A jet stream has also been observed in the upper stratosphere at high latitudes in the fall. It meanders around the hemisphere, much like the tropospheric jets, but its curves are generally much flatter.

On several occasions during the past 15 years, sudden, intense warmings have been observed in the upper stratosphere during midwinter. Temperature increases in the polar regions of 50°C or more have occurred during a period of ten days or less. When it happens, the normal westerly circumpolar flow is disrupted, with a high-pressure area extending over the polar regions. The origin of these explosive warmings and what effect, if any, they have on the weather of the troposphere, are still unknown.

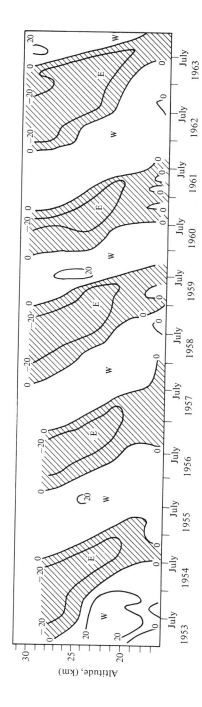

FIG. 5-6. The biennial cycle of the zonal (west to east) components of the wind in the equatorial lower stratosphere. (Canton Island; speed in ms^{-1}; courtesy of U.S. Navy Weather Research Facility.)

143

Cyclones and Anticyclones

Embedded in the great circumpolar vortex of the general circulation are the cyclones and anticyclones of middle and high latitudes. These smaller-scale vortices tend to be masked by the averaging process used to analyze the general circulation, but an inspection of the flow for an individual day reveals their existence. These smaller features of the flow can be seen, for example, in Figure 5-7. Compared to the size of the circumpolar whirl (about 10,000 km in diameter), the cyclonic and anticyclonic eddies range from several hundred to perhaps 3000 km in diameter.

These large cyclonic and anticyclonic eddies, that move in a general easterly direction around each hemisphere, dominate the flow over much of the earth between about 30° and 75° latitude (Figure 5-7). As was mentioned earlier, they are more significant transporters of heat and momentum between low and high latitudes than is the mean meridional (south-north) flow of the general circulation. The formation of these eddies over middle and high latitudes can actually be demonstrated by laboratory experiment: A round shallow pan is filled with water to simulate the thin atmospheric shell, and aluminum dust or other tracer material is placed on the surface of the water so that the motions can be observed. Then the pan is slowly rotated (about 4 rpm), while at the same time being heated at the rim to simulate the heating at the equator, and cooled at the center to represent the cooling at the pole. After a short time, a circulation pattern will be observed as shown in Figure 5-8. Note the similarity between the patterns of Figures 5-7 and 5-8.

The cyclonic whirls are the "storms" of middle latitudes. In the temperate latitudes they produce much of the winter precipitation. Around their low-pressure centers, the air circulates in a counterclockwise direction in the Northern Hemisphere and in a clockwise direction in the Southern Hemisphere. The masses of air that circulate around them are generally heterogeneous with respect to temperature and moisture, having come from different geographical areas; as a result, there exist sharp transition zones separating warm, moist air from cold, dry masses. These storm systems go through a complex life cycle which will be discussed in more detail in Chapter 6.

Winds in the anticyclonic circulations blow clockwise around their high-pressure centers in the Northern Hemisphere, and counterclockwise in the Southern Hemisphere. Within these whirls, the air is slowly subsiding at the rate of 10–15 cm/sec and "fair weather" gen-

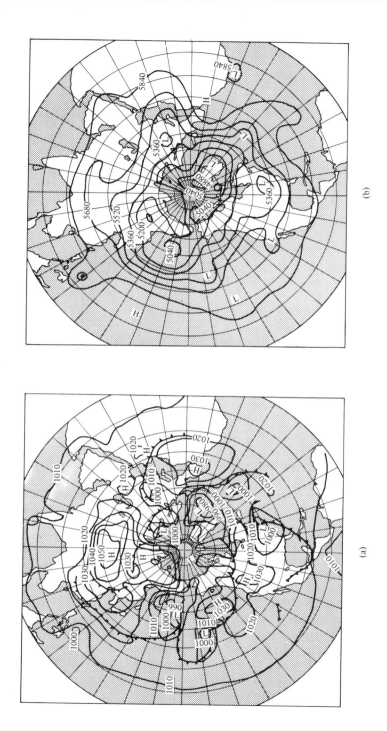

(b)

(a)

FIG. 5-7. Patterns of flow on December 29, 1959. (a) At sea level; (b) At 500 mb.

145

FIG. 5-8. Patterns of flow in a rotating pan of water. (After Fultz.)

erally prevails. The air masses of which they are composed are generally homogeneous with respect to temperature and moisture.

The Monsoon

Generally somewhat smaller in size than the cyclone or anticyclone is a thermal circulation called the *monsoon*. Its name, derived from the Arabic word for season, denotes its seasonal nature. In principle, the monsoon circulation is an excellent example of the thermal cell as can be seen by comparing Figure 5-9 with Figure 4-15. During the summer, when the land is warmer than the ocean, the air at the earth's surface flows onshore toward the land; during the winter, when the land is colder than the ocean, the flow is reversed.

In the regions where the monsoon circulation is especially well developed, seasonal precipitation is closely linked to the onshore and

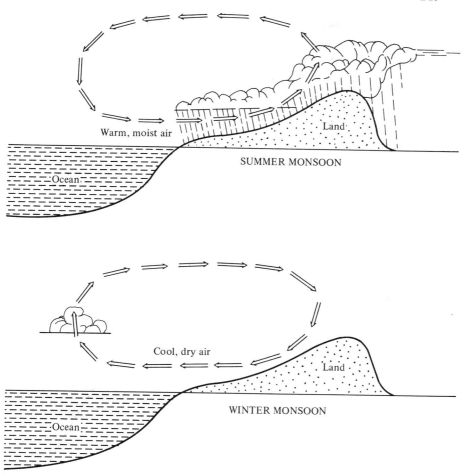

FIG. 5-9. The monsoon circulation. (Compare with Figure 4-15.)

offshore flow. The summer monsoon, flowing onshore, brings moist oceanic air to the land. The air is forced to rise over the land, producing cloudiness and eventual precipitation. But the winter monsoon, with the air subsiding and streaming offshore from the land to the ocean, is much less likely to result in precipitation.

The most intense monsoons are those produced by the large Asian land mass. In fact, the climate of southern Asia is largely determined by the monsoon. In summer, southerly winds over the northern Indian Ocean and a counterflow from the Arabian Sea converge over northern

India to deposit the heaviest rainfall in the world along the southern Himalayan slopes. Cherrapunji, which is located on these slopes at an elevation of about 1360 m, receives an average of about 11 m (36 ft!) of rain each year. Most of this rain is associated with the summer monsoon.

The summer monsoon is of great significance to agriculture in southern Asia, for the precipitation it brings largely determines the success of the crops in that region. However, the time of its beginning and its distribution, intensity, and duration vary greatly from year to year. During years of deficient monsoon rains, there are serious short-ages of food in countries such as India and Pakistan. Meteorologists have conducted intensive investigations of the causes for these fluctua-tions, but they still cannot be forecast with any degree of certainty.

During the winter monsoon, dry continental air flows offshore from the Asian land mass toward the ocean. Since the air is initially dry, it does not produce rainfall over the continent. However, the air soon picks up moisture as it travels over the warm ocean. Then, as it reaches the offshore islands; i.e., Japan, Taiwan, and the Philippines, heavy rain may fall on the western mountain slopes of these countries.

The continent of North America also experiences monsoon circula-tions, although these are not nearly as strong as those in Asia. More-over, most North American monsoons tend to be obscured by the effects of migratory cyclones and fronts.° The most noticeable monsoon occurs over the southwestern United States where, in summer, the hot interior draws in moist tropical air from the Gulf of Mexico and Carib-bean Sea. The summer thunderstorms over the arid highlands of New Mexico and Arizona can be attributed partially to the moist air from the Gulf of Mexico, forced inland by the monsoon circulation.

Land and Sea Breezes

The coastline in most regions is a boundary which reflects a sharp difference between the thermal characteristics of the ocean and the land. Because it is constantly being stirred by ocean currents, as well as for other reasons (see p. 85), the sea is a gigantic reservoir of heat. Thus the ocean, as well as the air immediately adjacent to its surface, rarely exhibits a change of more than 2°C between day and night. The land, on the other hand, heats up and cools off quickly between day and night. A diurnal temperature range of 20°C over dry land during

---

° One circulation may mask the presence of another. Recall that in the discus-sion on scales of motion it was mentioned that various circulations are superimposed on one another, like "wheels within wheels."

the summer is not unusual in many areas of the world. As a result, large temperature differences can occasionally develop across the coastline—during the day, the land will usually be warmer than the sea while at night this temperature gradient will be reversed.

These land-sea temperature differences lead to the creation of a thermal circulation in which there is a daytime flow from sea to land known as the *sea breeze*, and a nighttime flow from land to sea called the *land breeze*. This phenomenon is similar to the monsoon, but instead of an annual cycle associated with the change in seasons, the circulation changes direction each day. In principle, the illustration of Figure 5-9 can be applied to the land and sea breeze merely by changing the caption "summer monsoon" to "daytime sea breeze," and "winter monsoon" to "nighttime land breeze." Note, however, that the sea breeze is only occasionally associated with rainfall, although it frequently induces cloudiness if the marine air is forced upward over the inland slopes.

The sea breeze may occur throughout the year in the tropics, but at higher latitudes it is mostly a summer phenomenon. It usually begins to develop three or four hours after sunrise, and reaches its peak during the early part of the afternoon. At its peak, the circulation cell will usually have extended both inland and seaward about 20 km, although it has been found to penetrate inland as much as 60 to 70 km. The entire circulation cell, including the upper seaward flow, is not normally more than 1 km deep, although in the tropics it may reach depths of as much as 3 or 4 km. The surface wind is usually gusty and constantly shifting in direction.

As the forward edge of the sea breeze passes over a point in its landward penetration, the relative humidity may increase by 40 per cent or more, while the temperature may decrease by 5°C or more in less than an hour. Sometimes fog or low stratus clouds will accompany the sea breeze. Where the ocean temperature just offshore is unusually cold, as along the coast of central California, fog or low stratus clouds may accompany the sea breeze. Along the coast of Peru, where the coastal water is extremely cold, the forward edge of the sea breeze is so sharply defined that the fog bank appears as a solid white wall.

As the land cools in the evening, the sea breeze decreases in intensity, and usually ceases entirely by late evening. The land breeze, which is much weaker than its daytime counterpart, normally begins shortly before midnight and reaches its maximum development near sunrise. While the land breeze tends to return the marine air again toward the ocean, it is usually associated with lower wind speeds and is of shorter duration than the sea breeze, so that not all of the ocean air is swept back to the sea. The result is generally a net influx of marine air over coastal areas which are affected by the land-sea breeze regime. This is a fortunate circumstance for dwellers in certain coastal cities, since it means that a portion of the air pollution generated by the city is constantly being swept inland with each fresh importation of sea air, and not all of it is returned by the land breeze.

The sea breeze also plays an important role in moderating the temperature along the seacoast, an influence which may be felt for many kilometers inland. The breeze created by lakes is generally much less intense, and has a smaller width and depth. Along the shores of the Great Lakes, for example, the inland penetration is usually not more than a few kilometers. However, it offers a welcome relief from the summer heat for residents who live close to the shore.

Mountain and Valley Winds

Along mountain slopes, there is also observed a thermal circulation that has a diurnal cycle. During the daytime, beginning in mid-morning and continuing until near sunset, the wind blows up the mountain slope from the valley below. This is called a *valley wind*. Beginning around midnight and continuing until shortly after sunrise, the wind blows down the mountain slope. This is the *mountain wind*. Figure 5-10 illustrates these winds.

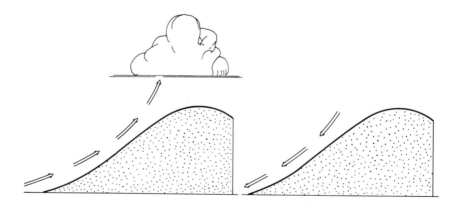

(a)   Daytime valley breeze   (b)   Nighttime mountain breeze

FIG. 5-10.   Mountain and valley breezes.

The mountain-valley circulation is produced because the air in contact with the slope is warmed by the sun during the daytime and therefore becomes less dense than the air at the same elevation over the valley. At night, the air in contact with the slope is cooled by radiation and becomes more dense than the air over the valley. As a result, the air over the slopes rises during the day and sinks at night. The mountain-valley winds are most pronounced on clear summer days, when the prevailing winds are weak. The intensity of the flow and its direction at any point also depends upon the degree of the slope, its orientation with respect to the sun's rays, and upon the configuration of the valley. Mountain and valley winds are best developed in deep valleys with mountain slopes facing the midday sun.

The rising air currents associated with the valley winds are a familiar phenomenon to every mountain climber, since they are frequently associated with cumulus clouds and showers over the mountain peaks. The height of these rising currents is usually between 100 and 200 m above the slopes, although when the air is unstable, the towering clouds they initiate may rise as high as 10,000 m.

Katabatic Winds

A mass of cold air over an elevated plateau during the winter tends to become more dense through radiative cooling, and will then drain down the slopes and into the valleys below. This is illustrated in Figure 5-11(a). The resulting downslope, drainage-type winds are called *katabatic winds.* Most are relatively gentle breezes, not exceeding 4 or 5 m/sec. Occasionally, however, the cold dense air may be set in motion by a migratory cyclone or anticyclone and the katabatic wind may then attain destructive violence.

One of the strongest of the katabatic winds occurs when air is cooled as it moves across the snow fields on a high plateau and cascades downward when it reaches the edge of the plateau. In some locations, such as along the coasts of Alaska, Greenland, and Norway, deep canyons or fjords confine the flow into a narrow channel so that the wind reaches very high speeds. Under these circumstances, the wind direction is little affected by forces other than that due to the pressure gradient—the wind may blow directly across the isobars from high to low pressure.

In some locations where katabatic winds are especially severe, they have been given local names. One of these is the *bora,* which sporadically brings cold air down from the Austrian Alps to the usually warm

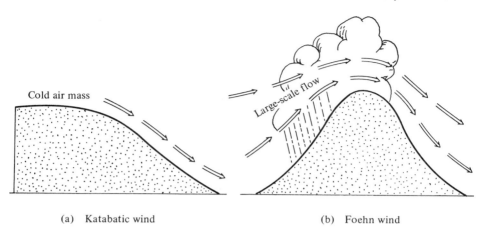

(a)   Katabatic wind   (b)   Foehn wind

FIG. 5-11.   Katabatic and foehn winds.

Adriatic Sea. Intermittent gusts as high as 50 or 60 m/sec have been recorded. A similar wind is the *mistral*, which occurs along the French coast of the Mediterranean Sea.

Foehn Winds

The *foehn wind* is a downslope flow of air which occurs in many mountainous areas, but it is not caused by the drainage of dense air. It occurs when the prevailing winds in warm, moist air are directed against a mountain. The forced ascent on the windward side, as illustrated in Figure 5-11(b), usually causes clouds to form. Frequently, precipitation will also occur. During most of this ascent, the air is cooled at the *moist* adiabatic rate (about 5°C/km), and by the time the air reaches the mountaintop, much of the moisture may have been removed. This means that the air at the top has absorbed the latent heat released by the condensation of the moisture it contained. As the air descends the lee slopes, it is warmed at the *dry* adiabatic rate (10°C/km). When it arrives at the bottom of the mountain, the air is warmer than at the same elevation on the windward side, having been heated by the latent heat of condensation. It is also drier, because it has lost some of its moisture through condensation and/or precipitation on the windward mountain slope.

Although the term foehn (from the German, *föhn*) originated in the Alps where these winds are especially frequent, some of the most dramatic types occur along the east slopes of the Rocky Mountains of North America. Here they are called *chinooks,* since they appeared to the early settlers to originate in the Indian territory of that name. The Indians themselves referred to the wind as the "snow-eater," because of the startling way in which large amounts of snow could be melted and evaporated by the warmth and dryness of the air. During the occurrence of a chinook, it is not unusual for a two-foot layer of snow to disappear in one day. At the same time, the temperature may rise as much as 30°C (54°F) in a few hours.

A strong foehn wind can be extremely disagreeable. Because of the high temperatures and abnormally low humidity, it is not unusual for people to feel irritable and out of sorts while the wind lasts. In fact, legend has it that the occurrence of such winds in the Sacramento Valley of California during the 1849 "gold rush" was sufficient justification for the many shooting affrays of that day. The wind also has an unfavorable effect on the mountain environment itself, bringing dangerous "fire weather" to the forests and woodlands of the region. The *Santa Ana* is a foehn-like, warm dry wind which blows from the Nevada plateau through the mountain passes of southern California. Its name derives from the Santa Ana Canyon where it attains especially high speeds, but it has occasionally caused damage to ships at anchor in the harbors around Los Angeles.

Other Local Winds

Particular geographical conditions of some areas of the world lead to certain characteristic winds, often in association with the migratory cyclones. A few examples are given in the following paragraphs.

Occasionally, in winter, cold air sweeps rapidly southward over the Great Plains of the United States behind a cyclone that has moved off toward the east. The Rocky Mountain barrier helps to channel the flow of cold air, and it has been known to reach as far south as Panama and perhaps even to Venezuela. In Texas, these strong, cold winds are known as Texas *northers* and in Central America as "nortes." In Kansas and Texas, the gush of cold air can cause the temperature to drop as much as 50°F in three or four hours. When the strong winds are accompanied by snow (usually, mostly fine, dry snow picked up from the ground), the weather condition is called a *blizzard.*

A similar phenomenon occurs in South America, except that the cold air comes from the south. When there is an intense cyclone over the South Atlantic, polar air streaks northward across the Argentine pampas (giving the phenomenon the name "pampero"), channeled by the high Andean range to the west, at speeds of 50 mph or more. Sometimes these surges of cold air are carried northward through Paraguay and Bolivia into the Amazon Basin right across the equator. Squalls and thunderstorms often accompany the pampero, but sometimes it is rainless or stirs up great dust storms.

There are similar cold winds of other regions: the *buran* of Siberia, the *burga* of Alaska, the *boulbie* of southern France. Nearer the tropics, there are the *levanter* of southern Spain and the *harmattan* of the west coast of Africa. The *sumatra* is a wind which accompanies the rain squalls of the coasts of Sumatra and the Malay peninsula during the monsoon season.

Warm winds are also given local names. The *sirocco* is a hot, dry, dust-laden wind which originates in the Sahara desert and blows northward across the Mediterranean in advance of a cyclone moving eastward. As the hot air crosses the Mediterranean, it picks up a great deal of moisture and causes very muggy, uncomfortable weather if it reaches Malta, Sicily, and southern Italy. Farther east, the *khamsin* blows from the Libyan desert northward across Egypt. Throughout the desert areas of Africa and southwest Asia, the *simoon* brings hot, dry weather and occasional disagreeable sandstorms to that region.

5-3. Thunderstorms

As we have already pointed out, vertical motion in the atmosphere is the key to many of the characteristics of weather. Upward motion results in expansion, cooling, and eventual condensation of the water vapor in a stream of air; the release of latent heat is often an important factor in accelerating the convection by increasing the buoyancy (instability) of the air. Downward motion results in compression, warming, and therefore an increase in the air's capacity for water vapor. We have seen, also, that convective patterns come in a large variety of sizes, the smaller ones nesting within the larger ones. In general, the maximum vertical velocity observed is inversely proportional to the size of the circulation: The large patterns have relatively feeble vertical motion while many of the small circulations have vertical motion equal to that in the horizontal.

Cloud types are closely related to the strength of the vertical motion. Stratified clouds, which sometimes extend unbroken over thousands of square miles, occur in gently ascending air (almost always less than 20 cm/sec; see Appendix 2). Cumuliform clouds, on the other hand, occur as isolated cloud masses (individual elements rarely cover more than 75 km²) and contain within them upward motions as strong as 35 m/sec. Stratified clouds form in air in which the buoyancy forces are weak or even oppose vertical motion above the thin layer in which the clouds form. For example, a wind blowing up a mountain slope may lead to condensation, but a temperature inversion may prevent vertical development of the clouds. Cumuliform clouds—those with great vertical development—are associated with instability.

The convective ascent of air in cumuliform clouds appears to occur in bursts or bubbles, somewhat like those that form in boiling water. Each successive bubble, having dimensions of up to a few kilometers in the horizontal and a few hundred meters in the vertical, rises, expands, and cools; but, as the bubble ascends through the atmosphere it is "eroded" or mixed with the surroundings, gradually losing its identity (Figure 5-12). In this way, puffs of cumulus may form, gradually disappear, and perhaps be replaced by new puffs. Horizontal winds may carry each bubble downstream from the surface point where it was created. Strong turbulent flow tends to cause more rapid erosion.

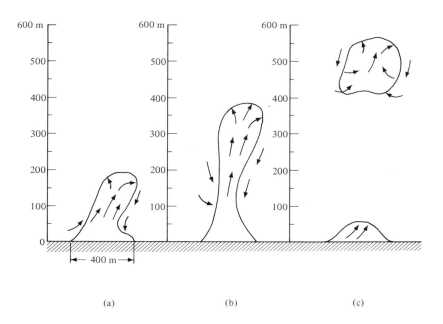

(a)   (b)   (c)

FIG. 5-12.   Development of convective bubbles producing a "thermal." (Note that the vertical scale is exaggerated.)

However, if the horizontal winds and turbulent mixing are not too great, previous bubbles may still remain while new bubbles are being formed. Then, each subsequent bubble will be able to ascend to ever greater heights, causing the cumulus cloud to develop vertically. Of course, an air bubble cannot rise without some other air descending to replace it, since there can be no vacuum.* With increased instability, enhanced by the release of latent heat within the cloud, the "percolation" becomes more continuous, and a well-defined thermal circulation, such as occurs in the towering cumulonimbus of a thunderstorm cell, may develop. The rapidity with which such a convective cell can develop is illustrated in the photographs of Figure 4-17. Of the many individual cumulus convective cells appearing on the horizon in the first photograph of Figure 4-17, one mushroomed vertically into a mature storm in only 18 minutes. It is not usual that isolated convective "cells" of this sort can be identified; normally, there is a tendency for adjacent cells to develop and join together. Frequently, there are great masses or lines of thunderstorms extending over 80 kilometers or more, but a single "cell" has a diameter of about 8 kilometers.

Studies have shown that there are three characteristic stages in the life cycle of a thunderstorm cell. These are illustrated in Figure 5-13. The initial, *cumulus* stage usually lasts for about 15 minutes. During this period, the cell grows laterally from 1 or 2 miles in diameter to 5 or 6 miles, and vertically to 25,000 or 30,000 ft. Note from Figure 5-13(a) that the updraft is strongest (about 20 mph) near the top of the cloud. Air is entering the cloud through the sides of the cloud at all levels. The upward motion is actually greater than the horizontal speed, which is the reverse of what is found in larger-scale atmospheric circulations.

The *mature* stage [Figure 5-13(b)] begins when rain falls out of the cloud base, and usually lasts for 15–30 minutes. During this stage, the size of drops and ice crystals in the clouds grows so large that the updrafts can no longer support them and they begin to fall as large drops or hail. The frictional drag of the precipitation gradually slows the updraft, and, in one part of the cell, a strong downward motion develops because of the precipitation and "entrainment" of cooler air

---

* Columns of rising air are sometimes called *thermals*. Glider pilots learn to seek out these thermals and then try to remain within their boundaries so they can be carried upward. All around the thermal, there is a compensating downward flow of air. Thermal upcurrents originating at the ground typically have a width that is about half to a third of the height above the surface. The updrafts generally range from 1 to 5 meters per second, while the downdrafts have velocities up to 10 cm/sec.

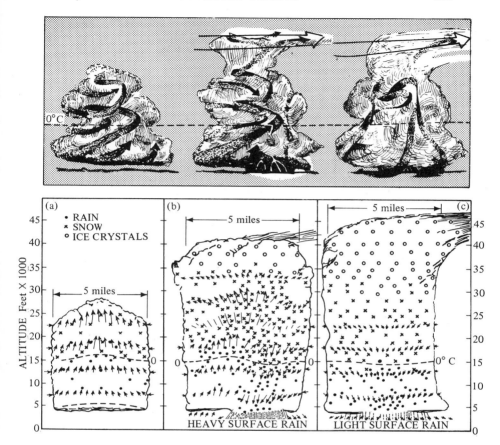

FIG. 5-13. Life cycle of a typical cumulonimbus cell. (Upper) Pictorial 3-dimensional representation of flow patterns. (Lower) Schematic 2-dimensional representation that shows the relative wind velocities, temperature, and distribution of liquid and solid water.

from outside the cloud. Near the center and top of the cloud, upward motion is still strong, however; speeds as high as 70 mph have been observed. Note also the strong outflow below the base of the cloud. When this downdraft meets the ground, it spreads away from the thunderstorm. It is for this reason that gusty, cool winds usually precede the actual arrival of a thunderstorm.

The mature stage is the most intense period of the thunderstorm. Lightning is most frequent during this period, turbulence is most severe, and hail, if present, is most often found in this stage. The cloud reaches its greatest vertical development near the end of this stage, usually reaching above 40,000 ft and sometimes penetrating the tropopause to altitudes greater than 60,000 ft.

The final or *dissipating* stage begins when the downdraft has spread over the entire cell. With the updraft cut off, the rate of precipitation diminishes and so the downdrafts are also gradually subdued. Finally, the last flashes of lightning fade away and the cloud begins to dissolve, perhaps persisting for a while in a stratified form.

Thunderstorms generally occur within moist, warm (maritime tropical) air masses that have become unstable either through surface heating or forced ascent over mountains or fronts. In the United States, practically the only source region of this air mass is the Gulf of Mexico and Caribbean. Note how the geographic pattern of thunderstorm incidence shown in Figure 5-14 is correlated with both the distance from the source region and topography.

A thunderstorm is, as the name implies, a storm accompanied by thunder and, therefore, lightning. As Benjamin Franklin demonstrated in 1750, lightning discharges are giant electrical sparks. Cumulonimbus clouds, therefore, are great natural electrical generators. Like man-

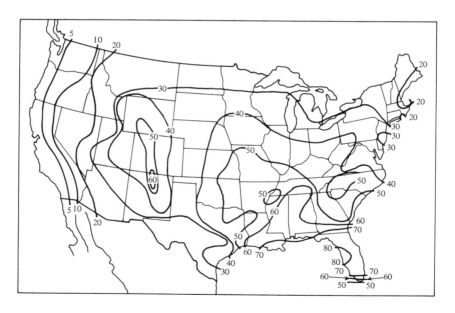

FIG. 5-14.   Average annual number of days with thunderstorms.

made machines, such as batteries, the cloud produces "poles" of concentrations of positive and negative electricity.

An important question that is still not adequately answered is how the powerful convective currents in these clouds produce the electrical charge and then separate the positive from the negative electricity. The lower part of a thundercloud has a concentration of negative charge, while the upper part is largely positive. The process that produces and separates charges must involve the water and ice particles in clouds. Some suggested processes are: (1) Friction between the ice particles formed near the top of such clouds would cause the ice to become negatively charged. Large ice crystals that fall would then carry negative electricity downward, leaving the upper portions positive. (2) Water droplets, when they form, tend to attract negative ions. (Ions are molecules that have become charged through the loss or gain of an electron. In the atmosphere, ions are produced by the radiations from radioactive material in the soil, by cosmic rays from the sun, and by combustion, friction, and splitting of water drops in sprays.) (3) If a cloud already has a predominance of positive charge near the top and negative near the bottom, so that an electrical potential exists, then any drop will tend to distribute its internal charge so that the bottom portion of the drop is positive and the upper portion is negative. If a current of air is sweeping upward past the drop, negative ions will be captured by the bottoms of the drops (which face the air current) more readily than will the positive ions. The rising air currents will therefore arrive near the cloud top with their negative ions depleted, or, in other words, with a positive charge.

Regardless of how the thundercloud does it, the fact remains that enormous potential differences are generated within clouds and between clouds and ground. Just before a discharge, the electrical potential gradient is of the order of 3000 volts per centimeter and potential differences between the extremities of flashes reach hundreds of millions of volts. A typical thunderstorm dissipates electrical energy at an average rate of about a million kilowatts.

Special photographic techniques have shown that individual lightning discharges actually consist of multiple strokes, each lasting about 0.0002 sec, with about 0.0001 sec between successive strokes. The air along the lightning channel is heated momentarily to between 15,000°C and 30,000°C (compared to the sun's surface temperature of about 6000°C); this causes a very rapid expansion of air, which in turn results in the deep sound called *thunder*. The rumbling of thunder is due to the fact that the sound is generated over a long discharge path, so that sound waves travel over many different paths to the observer, and much of the sound is reflected. The approximate distance of a thunderstorm can be computed by noting the time elapsed between a flash of lightning and the arrival of the sound wave by using the average speed of sound (330 m/sec, or 1080 ft/sec). The sequence of events during a lightning stroke is presented in Figure 5-15. The old proverb that lightning does not strike twice in the same place is, of course, untrue.

FIG. 5-15.   Life cycle of a lightning stroke. (a) Pilot leader starts a conductive channel toward ground; (b) Step leaders from cloud move downward for short intervals; (c) Streamers from ground meet leaders; (d) Return stroke from ground illuminates branches. Main stroke is followed by sequence of dart leaders and returns until potential is reduced. Total elapsed time: about one second. (After ESSA.)

Tall towers and buildings are repeatedly struck by lightning; Franklin's lightning rod protects such structures by providing a low-resistance conductor of the electrical current to the ground. One should always avoid being near an isolated, high target during a thunderstorm. Do not get caught under a tree or in an open golf course during a storm. Any open space where you may be the highest object extending above the surface is dangerous; you then become a "lightning rod."

PROBLEMS

Problems marked with an asterisk (°) are the most challenging.

1. Since the sun continuously provides the energy which drives the general circulation, why are the average winds of the earth not continually speeding up?

2. Describe the dominant features of the mean sea level pressure maps of winter and summer (Figure 5-2) in terms of positions and intensities of high and low centers and changes between winter and summer. Using tracing paper, plot the expected mean wind vector (length proportional to speed) at intervals of 10° latitude and longitude, both for summer and winter. What are the significant differences in the air flow over North America between the two seasons? Estimate the highest and lowest mean sea level pressures in January and in July. Using dividers or a compass, compute the greatest horizontal pressure gradient over a distance of 5° of latitude (555 km), both in summer and winter. Compare these mean pressure gradients in the horizontal with that found in the vertical near sea level.

3. Considering the surface winds of the general circulation, by what general route would sailing ships travel from London to New York? By what route would they return? Which of these two routes would be the shorter distance?

4. The normal flying time from New York to San Francisco is greater than the flying time from San Francisco to New York. Why?

5. Since the northeast trade winds prevail over most of the Northern Hemisphere between latitudes 10°N and 25°N, why are the winds over southeast Asia from the south during the summer?

6. As air crosses the equator, it tends to travel in a straight line rather than a curved path as it does at higher latitudes. Why?

7. If the air at the equator were not moving when viewed by an observer on Mars, what would be the direction and speed of the wind as measured by an observer on the earth?

8. In a given location, why is the sea breeze usually stronger than the associated land breeze?

9. Along the coastal hills of California, the sea breeze usually produces fog or low clouds on the windward side. Why? Although the wind blows directly across the hills, why do the fog and clouds usually dissipate on the leeward side?

10. An examination of the normal sea level pressure distribution for July shows a significant low-pressure system located over the area of the Mojave Desert, yet almost no precipitation falls in that region during the summer. Why?

11. On the average, the annual temperature range in the Northern Hemisphere is much greater than in the Southern Hemisphere. Why?

*12. Compute the height of the base of a cumulus cloud formed by a thermal, if the surface temperature and dew point are 85°F and 49°F, respectively.

6

Atmospheric Motions: Vortices

6-1. Rotational Motion

The more violent aspects of weather are associated with cyclonic rotating whirlpools of air called *vortices*. There are three principal types of *vortices* in the atmosphere: The *wave cyclone* of middle and high latitudes is the largest weather-producing vortex, but it is usually not the most violent. Typically, it has a diameter of about 2000 km, and the wind near the earth's surface usually does not exceed 70 km/hr (45 mph). Smaller in size (average diameter about 700 km or 450 miles) and much more destructive is the *tropical cyclone*. The maximum surface wind speed is sometimes more than 200 km/hr (125 mph). Fortunately, the tropi-

163

cal cyclone spends most of its life on the oceans where it does little harm. The smallest vortex, but the one with most powerful punch, is the *tornado*. The intense rotation of this vortex is confined normally to a diameter of a kilometer or less, but its wind speed can reach 300 km/hr (200 mph).

The extratropical wave cyclone is the best understood of these three vortices. By far the most common storm, it is found principally in the temperate latitudes and is large enough so that many of its characteristics can be ascertained without a great number of meteorological observations. Much less is known about the formation and structure of the more violent tropical cyclones and tornadoes. In the case of tropical cyclones, which form over the oceans and spend most of their lifetimes there, reports are too scanty to properly pin down their dynamics. Although tornadoes occur principally over continents, they are so small and have such a short lifetime (rarely longer than one hour) that they usually fall between the observation points of the standard land network of weather stations.

All three of these strongly rotating vortices are cyclonic; i.e., low pressure lies at the axis of rotation of each. The reason for this can be understood by recalling the discussions of the forces producing and affecting atmospheric motions (Chapter 4). When air converges toward a point, as it does in the case of a low-pressure center, its speed of rotation is increased. This results in a greater centrifugal force (and Coriolis force, away from the equator), both of which will be at least partially opposed to the pressure gradient force. There is no upper limit to the rotation speed since it can increase until these forces are in balance; the strength of the pressure gradient force (and friction) determine how fast the air will move. In contrast, when air is diverging from a high-pressure center (anticyclone), the centrifugal force resulting from rotation will have a component that acts in the *same* direction as the pressure gradient force. Since the centrifugal force increases with the square of the speed, anticyclonic rotation speed is strictly limited because the combined pressure gradient and centrifugal forces exceed any inward force (such as the Coriolis), and a balance of forces can only be achieved at a relatively small rotation speed.

Vorticity*

Fluids typically do not move in straight lines, but rather along curved paths. The rotation or spin of fluid elements is such a common

---

* This section is useful for a more basic understanding of the creation of vortices, but is not essential. Some theories for the development of vortices, such as the one explaining hurricane formation, are fundamentally explained in terms of "vorticity."

property that it is useful to introduce a measure of the spin as part of the description of the flow characteristics. This spin is referred to as *vorticity*. Each particle of air in the atmosphere can thus be assigned a value of vorticity, which would depend on the orientation of the axis around which it is rotating and the speed of rotation. For example, a particle moving around a low-pressure center in essentially a horizontal plane would be spinning around a vertical axis, and its vorticity would be proportional to the magnitude of the angular velocity. A particle moving around a high-pressure center in a horizontal plane would also be spinning around a vertical axis, but in the opposite sense.

The vorticity or spin of a particle can only change if there is a net tangential force acting on its surface (i.e., a *torque*) or if the distribution of the mass within the particle changes. For example, if the disk of air in Figure 4-20 were spinning in a counterclockwise direction around a vertical axis, convergence (concentration of the mass near the axis of rotation) would increase the rate of rotation, while divergence would reduce the rotation rate. This is analogous, of course, to the skater who makes himself spin faster by pulling his arms in toward his body. Among the tangential forces or torques that cause particles to change their spin is that caused when pressure and density surfaces do not coincide. For example, in the discussion of thermal circulations (Figure 4-15), we related the distribution of pressure and density in a vertical plane with a circulation created around a horizontal axis.

Friction is another force that affects rotation in a fluid, just as it slows the turning of a wheel. But in a fluid, friction can actually induce rotation. For example, the slowing of the portion of a parcel that is in close proximity to the earth's surface will cause the parcel to spin around a horizontal axis that is moving with the mean flow of the parcel along the surface.

Regardless of the vorticity a parcel may have relative to the earth's surface, it always has the additional (counterclockwise) spin of the earth's surface. As shown in Chapter 4, when we discussed the Coriolis force, the spin through the local vertical axis varies from zero at the equator to a maximum at the poles. In the absence of torques and convergence or divergence, a parcel maintains its total vorticity; i.e., the sum of the earth's vorticity and that of the parcel relative to the earth's surface. Thus, if a parcel changes its latitude and if its total vorticity remains unchanged, its vorticity relative to the earth's surface must change. For example, if the total vorticity of a *poleward* moving parcel stays constant, its cyclonic spin *relative* to the earth's surface must decrease because the cyclonic spin of the earth's surface beneath the parcel is increasing. In other words, poleward-moving parcels tend to acquire anticyclonic spin while those moving toward the equator acquire cyclonic vorticity.

In summary, then, we see that there are many reasons for the rotational character of atmospheric motions. As might be expected, rotary motion of all sizes is found in the atmosphere, with axes of rotation aligned both vertically and horizontally. This chapter will discuss just a few of the more dramatic weather-producing vortices.

6-2.   Air Masses, Fronts, and Wave Cyclones

In the discussion of the general circulation, mention was made of the importance of the large, essentially horizontal eddies in producing an exchange of air between the polar regions and the tropics. These large whirls are most active in the middle latitudes, where they are the chief weather producers. Periodically (in some places, every few days in winter), the *wave cyclones*, as they are called by meteorologists, develop along the boundary between warm and cool streams of air, later sucking these streams of air toward their centers; in the process, they transport "cold" equatorward and "heat" poleward. The prognosis of the formation and development of these cyclones is one of the principal tasks of the short-range forecaster in the polar and temperate regions of the earth. Two concepts are important in the explanation of the formation of these giant storms: *air masses* and *fronts*.

Air Masses

The source of most of the troposphere's heat and moisture is the earth's surface. As a result, on the average, both the temperature and the moisture decrease with height. In the horizontal, too, there are gradients of temperature and moisture induced by latitudinal variations in insolation and by variations in surface properties. The average horizontal temperature gradients in January and July are evident from the isotherms of Figure 3-16. Note, however, that even on the average, the temperature gradient is not the same everywhere (the distance between consecutive isotherms varies from place to place). For example, in the Northern Hemisphere in January there is a belt of closely spaced isotherms that undulates around the hemisphere: from southern Alaska to the Great Lakes, then northeastward near Iceland into the North Atlantic and then through Central Asia.

On a day-to-day basis, these belts of isotherms are much thinner than they are on the average. There are narrow zones across which the temperature and moisture change abruptly with broad regions in which there are very gradual changes of these parameters. The existence of large volumes of air that are quite uniform in the horizontal is largely the result of geographic causes. When air remains for several days or weeks over a uniform surface of the earth over which the solar radiation received varies little, the air soon comes into equilibrium with the surface. Heat and moisture are carried between the surface and the air through radiation and vertical stirring, so that soon there will exist approximate homogeneity at levels equidistant from the surface.

Although there are no extensive surfaces on the earth that are completely uniform in terms of temperature and moisture, there are large areas, such as the tropical oceans, over which conditions change very gradually. Air circulating for several days over these surfaces does achieve approximate horizontal uniformity up to altitudes of several kilometers and sometimes up to the tropopause.

Such a huge body of air, extending over thousands of kilometers, is referred to as an *air mass*. It is defined as a volume of air within which the temperature and humidity change gradually in the horizontal; i.e., one which contains no sharp changes of temperature or humidity in the horizontal. Air masses are created principally within the anticyclonic flow of the subtropical and polar high-pressure belts. The air circulates slowly in these systems over surfaces of fairly uniform properties and gradually acquires thermal and moisture characteristics representative of these surfaces. For example, the air masses flowing around the semipermanent Atlantic anticyclone very quickly acquire the warmth and moisture of such water bodies as the Caribbean and Gulf of Mexico; they often extend to great altitudes in the troposphere. Cold air masses, such as those that form over the frozen surfaces of northern Canada in winter, take somewhat longer to form, but under fairly stagnant conditions horizontal homogeneity can exist to a 3- or 4-km depth.

Air masses are classified according to their geographic source region: *polar* or *tropical, maritime* or *continental*. The chief air masses that affect the weather of North America are continental polar (cP), maritime polar (mP), and maritime tropical (mT). The origin of continental polar air masses is northern Canada. In winter, the cP air mass is dry and stable before it moves from its source region, but when it moves southward over the United States it is heated from below and its stability decreases. The portion that traverses the Great Lakes picks up moisture, which frequently results in snow showers along the eastern shores of the lakes and in the Appalachian Mountains. Occasionally, this cP air may penetrate the Rocky Mountain range.

The maritime tropical air that affects the United States generally comes from the Gulf of Mexico. In winter, maritime polar air sweeping out of the Pacific is largely responsible for the winter rains of the west coast of the United States. As it strikes the coastal range and then the Rockies, the forced lifting causes heavy rain and snow over these barriers.

After it has left its source region, an air mass can be further characterized by its temperature relative to the surface over which it is traveling. An air mass is said to be *cold* if it is colder than the underlying surface and *warm* if it is warmer than the surface. A cold air mass will be heated from below, so that the lapse rate will increase, while a warm air mass will lose heat to the underlying surface and its lapse rate will decrease (it will become more stable).

Fronts

A sharp contrast of temperature and humidity may exist across the boundary that separates two air masses of differing properties. Such a boundary, where air masses "clash," is called a *frontal zone* or, more commonly, a *front*. The name "front" was coined by the Norwegian meteorologists who first developed the polar front theory during World War I, possibly because the oscillations of the boundary, with periodic flareups of weather along it, reminded them of the long battle line in Europe with its intermittent activity.

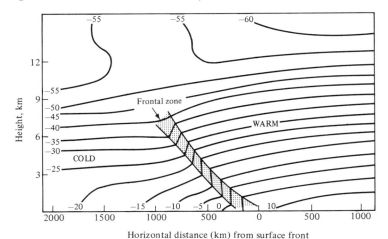

FIG. 6-1.   Vertical cross section of a front. (North-south cut, North America, on a winter day; isotherms in °C.)

The front separating two air masses slopes upward over the cold, denser air. This is illustrated in Figure 6-1, a typical vertical cross section of fronts over the center of the North American continent. Note that the vertical scale is greatly exaggerated. The average slope of fronts is only about 1:150, ranging from as little as 1:250 to as steep as 1:50. The width of the front—the transition zone between air masses— is usually about 50–100 km, but on the scale of distances that we are considering, such a width is closely approximated by the thickness of a heavy line drawn on a weather map.

The boundary between the warm and cold air masses must always slope upward over the cold air. This is because the cold air is a denser fluid. (Imagine two fluids such as water and oil, side by side, separated by a partition. If the partition is removed, the heavier water will slide beneath the oil.) Now if either the warm air is moving against the wedge of cold air or the wedge is pushing under the warm air, there will be forced lifting. In either case, cooling due to expansion may lead to condensation and possibly precipitation over the frontal surface.

Wave Cyclones

The weather pattern associated with migratory cyclonic depressions of the middle latitudes has been known for about 80 years. The polar front model relates the formation and maturation of these storms with undulations of the frontal boundary. A front separates air masses of different densities. The air masses flowing side by side may develop zones of strong wind "shear" between them; i.e., the currents of air on both sides of the boundary may have different velocities. In such an event, both air currents tend to acquire a spin; i.e., angular velocity. (This is like giving a stick a rotation by placing it between the palms of your hands and then moving your hands in opposite directions.) Under such conditions, part of the frontal boundary will begin to turn about some point along the front, forming a wave in the surface. Figure 6-2 shows the early stages of the development, looking down on the surface of the earth (Northern Hemisphere); the air masses and frontal boundary are rotating in a counterclockwise direction around the "apex" of the wave.

The complete theory of the development of this wave is rather complex. It is really a combination of many types of undulations, such as "gravity waves"—those that you see in water when you disturb the surface by dropping a stone in it; shearing waves, such as those that form when wind blows along a water surface; and inertial waves (earth's rotational effect, as illustrated in Figure 4-7). It turns out that

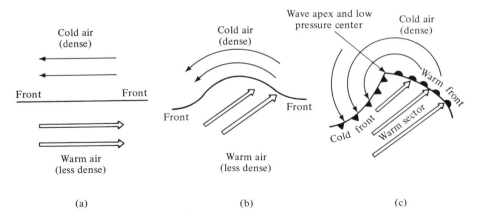

FIG. 6-2. Genesis and early development of a wave cyclone. (Northern Hemisphere; arrows depict air flow.)

for certain wavelengths (the overall length of the disturbance) these composite oscillations are unstable and the wave grows in amplitude and circulation intensity. Waves having lengths between about 600 and 3000 km are unstable, and these are the ones that experience development beyond stage (b) of Figure 6-2.

As the wave develops, low pressure forms at its apex [Figure 6-2 (c)] and both the warm and cold currents move in a cyclonic pattern around it. To the left (in the figure) of the apex, the front is advancing toward the warm air and this segment of the front is called the *cold front;* to the right of the apex, the front is receding from the warm air and so this segment is called the *warm front.* The warm air between the fronts is known as the *warm sector.*

Figure 6-3 represents an idealized wave cyclone—the view looking downward on the earth shown in the center and vertical cross sections taken a little south (bottom drawing) and a little north (top) of the apex. Imagine the entire system moving toward the right (eastward), as such systems normally do. If you were standing to the east and south of the apex, ahead of the warm front, the first sign of the approaching system would be high cirrus clouds. As time goes on, the wisps of cirrus thicken to cirrostratus clouds; these often cause halos (rings around the sun or moon), a sure sign of rain within 24 hours, according to a well-known proverb. Gradually the clouds lower and thicken to altostratus. The pressure falls, and the wind increases and backs (changes direction in a counterclockwise direction) as the low center gets closer. The temperature begins to rise slowly as the frontal

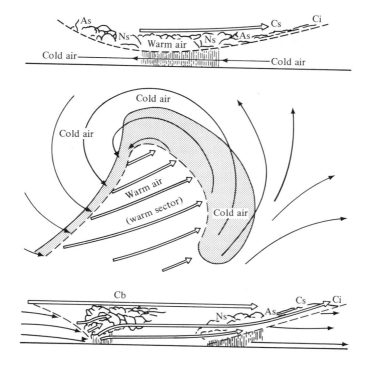

FIG. 6-3. The wave cyclone model. (After J. Bjerknes and H. Solberg.) Center drawing: horizontal plane view; Top: Vertical cross-sectional view just north of wave apex; Bottom: vertical cross-sectional view across warm sector. (For abbreviations of cloud type names, see p. 40; arrows depict air flow.)

transition zone approaches. Within 300 km of the surface position of the front, precipitation begins, either in the form of rain or snow. After the warm front passes, the precipitation stops, the wind veers (changes direction in a clockwise direction), and the pressure stops falling. Within the warm sector, the weather depends largely on the stability of the warm air mass and the surface over which it is moving; there may be showers or almost clear skies.

The type of weather accompanying the passage of the cold front depends on the sharpness of the front, its speed, and the stability of the air being forced aloft (typically, at speeds up to 1 m/sec). Usually, there are towering cumulus and showers along the forward edge of the front. Sometimes, especially in the midwestern United States during the spring, severe *squalls* precede the front. But in other cases, nimbo-

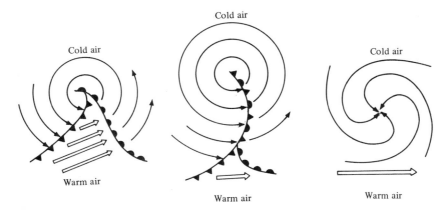

FIG. 6-4. Later stages in the development of a wave cyclone. (Northern Hemisphere.)

stratus and rain extend over a zone of 75–100 km. After the frontal passage, the wind veers sharply and the pressure begins to rise. Within a short distance behind the cold front, where the air is descending at a rate of up to 1 m/sec, the weather clears, the temperature begins to fall, and the visibility greatly improves.

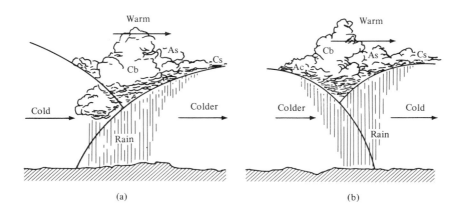

FIG. 6-5. Vertical cross section of occlusions. (a) Warm-front type; and (b) Cold-front type. (Arrows indicate displacement of air masses.)

The early genesis stages of the wave cyclone shown in Figure 6-2 normally take between 12 and 24 hours. Subsequent development of the wave, shown in Figure 6-4, takes an additional two or three days. As the wave breaks, the cold front begins to overtake the warm front. This process is called occlusion and the resulting boundary an occluded front. The vertical cross sections of Figure 6-5 illustrate that either the cold front can move up along the warm front (warm-front type occlusion) or it can force itself under the warm front (cold-front type). The occluded front is the boundary that separates the two cold air masses.

The shorter wave disturbances—those less than 600 km in wavelength—are generally damped out by friction. They usually do not last longer than 36 hours; the low center near the apex is weak, and any weather they produce is confined to a narrow zone along the frontal boundary. Wave cyclones usually occur in groups or "families" of between three and five, as illustrated in Figure 6-6.

During most of the wave cyclone's history it is in an occluded state. Gradually, the fronts begin to dissolve as the kinetic energy of

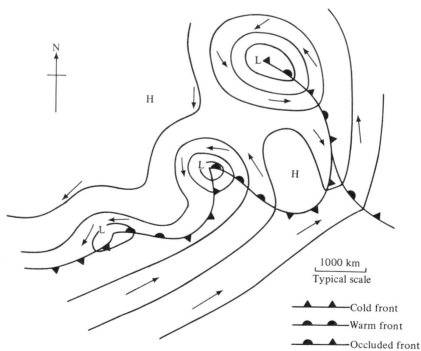

1000 km
Typical scale

Cold front
Warm front
Occluded front

FIG. 6-6. A "family" of wave cyclones. (Northern Hemisphere, arrows indicate air flow.)

the whirl is dissipated by friction and the winds subside. The maximum intensity of the wave cyclone, in terms of horizontal pressure gradient and wind velocity, normally occurs during the occlusion process. This happens because occlusion produces a redistribution of the air masses: The denser, cold air moves in at the surface of the system and less dense, warm air is forced aloft. This change in the distribution of mass within the eddy results in a loss of potential energy (more light air aloft, and more heavy air below) which reappears as kinetic energy— winds. Of course, the whirl is continuously being slowed by surface friction and thus losing some of its kinetic energy. When, in the last stages of the cyclone's history, there is little further readjustment of air masses and the supply of kinetic energy is cut off, friction gradually brings the giant eddy to a stop.

The net effect of the wave cyclone's history is to disrupt the initial air mass distribution. Part of the cold-air mass is swept to lower latitudes near the surface, while some of the warm-air mass is transported to higher latitudes aloft.

The sequence of events associated with wave cyclones—as described above—is, of course, an idealization. Few wave cyclones adhere closely to the model throughout their development. However, the model does serve as a guide in furthering understanding of these vortices. Since, on a typical winter day, about ten of these wave cyclones dot each hemisphere, they are the most common vortices in the atmosphere that are significant "weather producers."

6-3. Tropical Cyclones

Tropical cyclones, as their name implies, are cyclonic whirlpools that are formed over the tropics. In fact, they almost invariably form over the oceans in the latitudes between about 5° and 20° from the equator. They are spawned over all of the tropical oceans except the South Atlantic. Each area of the world has its own local name for this storm, the most common being: *hurricane* (North America), *typhoon* (Eastern Asia), *cyclone* (India), *willy willy* (Australia), *baguio* (China Sea).* Figure 6-7 gives the more common points of origin and paths of tropical cyclones in the world. In the discussion that follows we shall use the term, hurricane, in reference to these storms.

---

* By convention, these names are usually given only to those tropical cyclones in which the maximum wind speed is at least 74 mph.

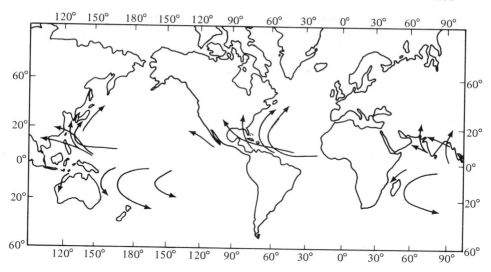

FIG. 6-7. Paths of tropical cyclones.

A bird's-eye view of a hurricane, taken from a weather satellite, is shown in Figure 6-8. Note how the bands of clouds spiral in a counterclockwise direction inward toward the center (Northern Hemisphere). From this photograph alone, one could hardly imagine the violence within it. Figure 6-9 presents a three-dimensional view of the cloud and air-circulation pattern. The illustrations of Figure 6-10 represent typical conditions observed along a radius of the three-dimensional view of Figure 6-9; since hurricanes are approximately circular in form, the conditions shown would be approximately the same along all radii extending from the center of the storm. The weather that normally can be expected during the approach and passage of a hurricane can be determined by imagining the observer moving slowly (usually 10–15 mph) from the outer edge to the center (right to left in Figure 6-10) and then out to the edge again.

High clouds, which are not too common over the tropical oceans, usually appear 200–300 miles in advance of the hurricane. The pressure begins to fall slowly and the winds begin to pick up above the normal 10–20 mph of the trade winds. Within 200 miles of the center, the winds reach gale force (about 30 mph), steadily increasing in speed, and the pressure begins to fall off a little more rapidly. By the time the observer is within 100 miles of the center, the winds will be 50

FIG. 6-8.   Satellite photograph of a hurricane.

mph or more, the clouds will be low and menacing, and the pressure will be falling rapidly. Rain usually starts falling 60 or 70 miles from the center and increases in intensity until it is coming down in torrents at 20 or 30 miles from the center. Winds in this last zone may be as high as 125 mph.

EASTERLY TRADEWINDS

EYE

DESCENDING AIR

FIG. 6-9. Three-dimensional structure of a hurricane.

177

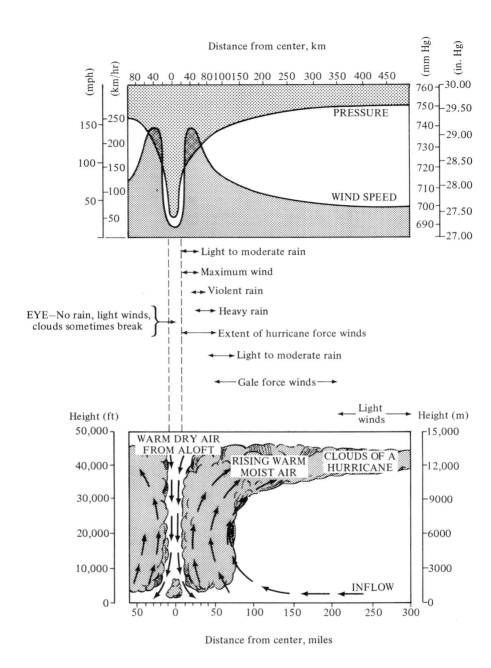

Distance from center, km

PRESSURE

WIND SPEED

→ Light to moderate rain

→ Maximum wind

←→ Violent rain

→ Heavy rain

EYE—No rain, light winds, clouds sometimes break

← Extent of hurricane force winds

← Light to moderate rain

← Gale force winds →

Light winds

Height (ft)

WARM DRY AIR FROM ALOFT

RISING WARM MOIST AIR

CLOUDS OF A HURRICANE

Height (m)

INFLOW

Distance from center, miles

FIG. 6-10. Vertical section of a hurricane and the associated patterns of wind, pressure, and rain.

If the center of the storm passes over the observer, he has a truly startling experience. This center is known as the *eye* of the hurricane. (The high-altitude photograph of Figure 6-11 would indicate that "eye" may not be a misnomer.) Quite abruptly, the winds decrease from 150 mph or more to less than 20 mph in a distance of 15 miles or less. (This distance corresponds to a time interval of less than an hour for the typical hurricane movement.) The rain stops completely, the clouds become thin, and the sun may shine through breaks. The clouds surrounding the eye appear as nearly vertical walls, extending from 2000 or 3000 ft to 40,000 ft or more. This is but a respite from the

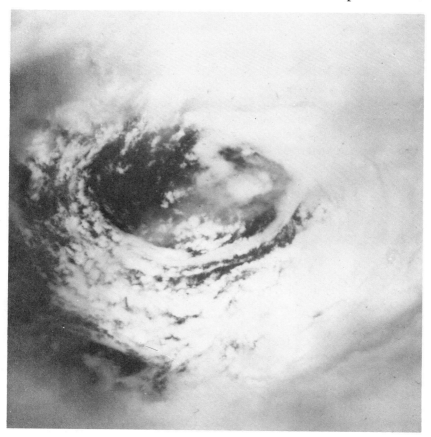

FIG. 6-11. The eye of hurricane Betsy photographed from an altitude of 11 miles at 1350 EDT, September 2, 1965, about 90 miles north of Grand Turk Island. (Courtesy of U.S. Air Force Weather Service.)

monster storm. Soon the other half of the "doughnut" will strike and the observer will experience weather conditions similar to those encountered before, except that they will occur in reverse order and the wind direction will be opposite. The average diameter of the hurricane eye is about 14 miles.

Hurricanes occur rather infrequently, compared to their brothers, the much larger wave cyclones of the middle and high latitudes. Table 6-1 gives estimates of the average number of tropical disturbances in various areas of the globe. Note that there are only about 40 per year in the Northern Hemisphere, making them less than a tenth as frequent as middle-latitude wave cyclones.

The West Indies hurricane season extends from June through November, although most hurricanes occur during August, September, and October. During the 81-year period of 1886 to 1966, 389 hurricanes (an average of 4.8 per year) were reported in the North Atlantic and adjoining waters, in addition to 259 other tropical cyclones that did not reach the hurricane intensity (officially, having wind speeds greater than 73 mph). About 4 per cent of these occurred in the month of June, 6 per cent in July, 29 per cent in August, 36 per cent in September, 19 per cent in October, and 3 per cent in November; there were only five hurricanes in the other six months of the year. The number of tropical cyclones per year has ranged from as few as two to as many as twenty-one. In the United States, it is customary to identify each season's hurricanes by giving them girls' names in alphabetical succes-

TABLE 6-1

Average Annual Frequency* of Tropical Cyclones

Northern Hemisphere		*Season of Occurrence*
Atlantic Ocean	8.0 (4.8)†	June–November
Pacific Ocean (off west coast of Mexico)	6.4 (2.5)	June–October
Pacific Ocean (west of 170E)	22.5 (16?)	May–November
Indian Ocean (Bay of Bengal)	6.0	May–November
Indian Ocean (Arabian Sea)	1.5	May–November
Southern Hemisphere		
Pacific Ocean (East coast of Australia)	3.1	December–April
Indian Ocean (Northwest Australia)	2.1	November–April
Indian Ocean (Madagascar to 90E)	5.1	November–May

* Statistics of areas outside the heavily traveled Atlantic are not considered to be reliable.

† Numbers in parentheses: those reaching hurricane intensity.

sion; thus, the first half-dozen Atlantic, Caribbean, and Gulf of Mexico storms of 1968 were called Abby, Brenda, Candy, Dolly, Edna, and Frances.

The average lifetime of a West Indies hurricane is nine days, although those occurring during August appear to be more durable, lasting for an average of twelve days. Hurricanes tend to move in the direction of the flow in which they are imbedded, much like an eddy in a river moves downstream. During their early stages in the Atlantic, while they are still well within the easterly winds, they tend to move toward the west or northwest. If they reach north of about 30° latitude before dissipating, they get caught by the prevailing west winds of the middle latitudes, and are swept toward the northeast. (See Figure 6-12.)

Hurricanes sometimes move in a very erratic fashion. For example, Hurricane Flora (October, 1963) meandered about over eastern Cuba for almost five days and Hurricane Betsy (September, 1965) started toward the northwest over the Bahamas and then passed through the Florida Strait, finally crashing into Louisiana. Although the average speed of hurricanes is about 12 mph, the speeds of individual storms are extremely variable; when they get caught up in the usual west winds north of 30° latitude, they frequently greatly accelerate, sometimes achieving a speed of over 50 mph. The variation in paths, speeds, and lifetimes of tropical cyclones is illustrated by the tracks shown in Figure 6-12.

The warm, moisture-laden air of the tropical oceans possesses an enormous capacity for heat energy, and it is estimated that most of the energy required to create and sustain a hurricane comes from what is released through condensation. The hurricane is a powerhouse of energy. Although only about 3 per cent of this latent heat is transformed into kinetic energy, hundreds of millions of tons of air are circulated at speeds of up to 125 mph or more. The average hurricane generates 300 to 400 billion killowatt-hours of energy per day, about 200 times the total electrical power produced in the United States. An average hurricane precipitates 10 to 20 billion tons of water each day.

A hurricane is an unusually well-organized, very large convection system that pumps great amounts of warm, moist air to high levels of the atmosphere at very rapid rates. The arrows in the vertical cross section of Figure 6-10 illustrate the overall convection pattern within a fully developed hurricane. Warm, moist air rises sharply in the ring between 10 and 50 miles of the center. Because of condensation, the temperature above about 1000 m within this ring rises sharply toward the "eye" of the hurricane; the eye is considerably warmer than the

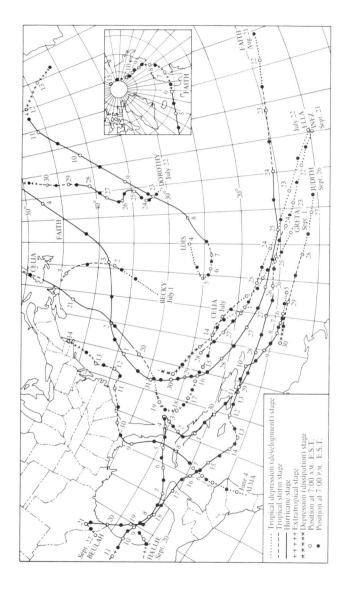

FIG. 6-12. Tracks of the tropical storms and hurricanes in 1966 and Hurricane Beulah (Sept. 5–22, 1967). (Solid lines designate those having hurricane intensity; dotted and dashed lines indicate tropical disturbance stage; hashed lines, dissipation stage. Positions are shown at 12-hour intervals, with dates given at 0700 EST.)

exterior of the hurricane (as much as 10°C or more at 10 km). As the air within this ring ascends, new air flows in toward the center from hundreds of miles away. If the air starts out with even a slight rotary motion (this is thought to be caused by the Coriolis force), it will spin faster and faster as it nears the center.

The circulation of a well-developed hurricane extends in the vertical to 14 or 15 km, almost to the level of the tropopause, although the intensity of the cyclonic circulation decreases with height. Typically, the peak velocity at 10 km is half what it is near the surface. In the lowest 3 km, there is a pronounced component of motion toward the center of the hurricane, causing convergence of the air and the ascending motion that leads to cloud formation (Figure 6-10). Above about 7½ km, and reaching a maximum rate at 12 km, the air flows with an outward component. Between these two layers (from 3 to 7½ km), there is little inflow or outflow. Frequently, in the uppermost layer, the air flow changes direction sharply as it moves outward from the center, and even acquires an anticyclonic rotation some 80 to 100 miles from the center.

The formation process of hurricanes is at yet not definitely established. Most meteorologists agree that some sort of "priming" is needed to start the flow of the hurricane. It has been suggested that wavelike motions in the trade winds that look something like Figure 6-13(a) may start the hurricane's development. Within a day or so, the circulation touched off by such a wave may take on the appearance of Figure 13(b) and, if it continues to develop, the flow may become a tight spiral, as illustrated in Figure 6-13(c). Or, it has been suggested that these vortices develop when the anvils of tropical cumulonimbus clouds spread out in the horizontal as they encounter the tropopause. The mechanism depends on the small-scale mixing processes that are presumed to occur in clouds. If the air particles within the anvil cloud are all initially rotating at uniform angular velocity around some point, O, as illustrated in Figure 6-14(a), the horizontal wind velocity will increase in proportion to the distance from the center of rotation. Individual elements are all, therefore, rotating around local axes, as is, for example, the element a as it moves to a'. Turbulent mixing within the anvil cloud acts to reduce the rotation of these individual elements; in order for this to occur, the velocity along radii must actually decrease with distance from the center [Figure 6-14(b)]. This is accomplished by a sharp increase in speed near the center with a decrease at the outer edges. (The element b is then not rotating about its central point.) Of course, the speed at the very center must be zero, so the peak actually occurs a short distance from the center. The result

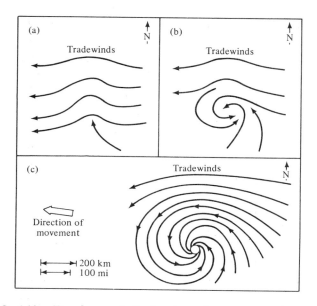

FIG. 6-13. Development of a hurricane from an easterly wave.

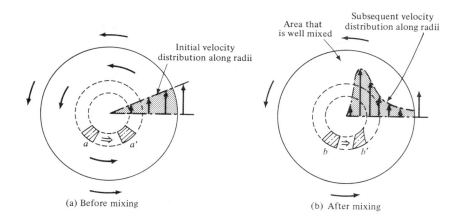

(a) Before mixing

(b) After mixing

FIG. 6-14. Proposed mechanism for initiation of hurricane circulations. (a) Initial velocity distribution before mixing; (b)
After mixing within an anvil cloud.

is a tremendous concentration of rotation close to the center; i.e., the
intense vortex that is the principal feature of the hurricane. The great

increase in kinetic energy is derived from the latent heat of condensation released within the cloud.

The greatest damage and loss of life during hurricanes results from flooding of coastal areas by the ocean surges and waves caused by the wind. The sea is in an agitated state hundreds or even thousands of miles from the storm center. When wind blows along a water surface, it exerts a frictional drag on the water that results in ripples or "waves." The wind drag increases with higher wind speed and so does the size of the waves generated. During a hurricane, air travels at high speed over long distances, producing waves of great height. The wave heights (vertical distance from crest to valley) often reach 35 ft, and sometimes exceed 50 ft in the zone of strong winds. As waves move out from under the winds that generated them, the crests decrease in height and become more regular in shape. Waves of similar height and length between consecutive crests tend to move in groups. These composite waves are known as *swells*, and they can travel thousands of miles from the generating area with little loss of energy. When these swells approach a coast, the varying depth to the ocean bottom and the irregularities of the coastline complicate the wave structure. Sometimes, very steep waves travel up estuaries, damaging vessels and piers. If these swells coincide with the normal high tide of the area, they may cause extensive flood damage.

However, the really damaging effects of the wind-churned ocean are not felt until the hurricane center is within a hundred miles or so of the coast. Rapid rises in the water level, known as *surges*, result from a piling up of water along the coast by the driving winds. Such "hills" of water can be 15 ft or more above normal sea level. With storm waves riding 20 ft or more above these mounds, large inland areas can be inundated. During a hurricane in 1900, Galveston, Texas, was flooded by just such a surge, which demolished the city and drowned about 5000 persons. In 1961, when Hurricane Carla struck the Gulf coast, a surge was predicted and the affected areas were evacuated beforehand, so that fewer than 50 lives were lost.

When a hurricane moves off the ocean onto land, much of its energy is cut off. In addition, the frictional drag that the surface of the earth exerts on the wind is greatly increased, the air speed is slowed, and the air takes a more direct path toward the low center. This more rapid inflow toward the center leads to the gradual dissipation or "filling" of the storm, but at the same time to somewhat heavier precipitation.

Although the genesis of tropical storms is still very difficult to predict and, once formed, their movement is quite erratic, the use of aircraft and radar allows the United States Weather Bureau to maintain a close watch on their path and to issue advance warnings of their approach. As a result, in recent years, there have been few deaths or injuries, even though property damage is high. The hurricane's effects are not all bad. In some areas of the world, much of the annual precipitation comes from tropical cyclones which sometimes break droughts.

6–4. Tornadoes

The name *tornado* is probably derived from the Spanish word "tornar," which means "to turn." A tornado is an intense cyclonic vortex in which the air spirals rapidly about a nearly vertical axis. Seen from a distance, it looks like a gray funnel or elephant's trunk extending downward from the base of a cumulonimbus cloud (Figure 6-15). Where this pendant cloud reaches the ground, great masses of dust and debris circle the lowest couple of hundred feet.

The winds associated with tornadoes are too strong to be withstood by the ordinary anemometer, so there are few reliable measurements. Estimates from damage to buildings and the impact force of flying objects indicate that speeds range generally between 100 and 300 mph, although it is possible that speeds up to 500 mph can occur. Such a wind necessitates a very strong pressure gradient. The pressure drop between the outside and inside of a tornado is usually of the order of 25 mb, but falls up to 200 mb have been observed.

The lengths of tornado paths average only about 6 km, but they are extremely erratic. Some touch ground along a path of only 200 or 300 m, while others hop and skip over tracks of hundreds of kilometers. (A tornado moved along a 300-mile path in Illinois and Indiana in a little more than 7 hours in 1917.) Some tornadoes hardly move, while a few have been known to travel at a speed up to 200 km/hr (125 mph). Some last only a fraction of a minute, while others persist for several hours; the average duration is less than 10 minutes. Most move toward the east or northeast (Northern Hemisphere), but every direction of movement has been observed.

During their brief lives, tornadoes can be very destructive. A building in the path of a tornado will certainly be badly damaged, if not destroyed. The cause of the damage to buildings is threefold: the enormous force exerted by the wind, the sudden pressure difference created between the interior and the exterior of the building, and the

FIG. 6-15.   Tornado. (Courtesy of ESSA.)

strong upward air currents. With a rapid pressure drop of 100 mb, the net outward pressure on the walls of a building could exceed 200 pounds per square foot, and buildings have been observed to literally explode. Wind pressure can easily reach several hundred pounds per square foot. And powerful updrafts may lift very heavy objects. Railroad cars have been lifted completely off their tracks by tornadoes. Many freak occurrences have been reported during tornadoes, such as showers of frogs that were sucked up from ponds miles away, the "defeathering" of chickens, straws driven through posts, and entire buildings carried for hundreds of meters.

Tornadoes occur infrequently. Although they have been observed in every part of the world outside of the extremely cold regions, they are most common over large continents, where strong horizontal temperature contrasts exist: in the United States east of the Rockies, in the southern and middle USSR, and in southern Australia. In the United States, there are about 200 per year,* mostly in the central plains states. Oklahoma, Kansas, Iowa, Arkansas, and East Texas have the highest frequency of tornadoes per unit area (Figure 6-16). They occur principally in the afternoon during the spring (mostly in April, May, and June in the United States), but can occur at any time during the day throughout the year. The places of highest tornado frequency have relatively low population density, so that tornadoes cause comparatively few deaths and injuries, despite their violence. The greatest destruction over the years has been in the Chicago area where a moderate incidence of tornadoes coincides with a high population density.

The formation mechanism of tornadoes is still somewhat obscure. They form generally in the vicinity of intense cold fronts and *squall lines* (moving lines of thunderstorms). Marked instability is an important factor, but such conditions often exist without tornadoes being produced. Evidently, there must be some special circumstances that lead to the sudden creation of a deep low-pressure center before it can be filled by the inflow of surrounding air. Probably, there is a combination of vertical instability, which provides the energy for the motion, and a mechanical impetus: strong shearing action of air currents in juxtaposition that creates the necessary spin. Once formed, strong convection action will sustain the vortex until the potential energy has been dissipated and friction destroys the whirl.

Tornadoes are often associated with hurricanes. For example, Hurricane Beulah (September, 1967) spawned 47 tornadoes in the vicinity of Brownsville, Texas, as it moved inland and gradually dissipated (Figure 6-12). Of the fifteen lives lost in Texas as a result of the hurricane, five were due to tornadoes, ten due to flooding.

Tornadoes occasionally form over warm water. Because of the high moisture content of the air, the funnels are heavily laden with water drops, so that they look somewhat like a stream of water pouring from the cloud base (Figure 6-17). They are called, for this reason,

---

* Because of their small size and short duration, many tornadoes probably go unreported. Over the past couple of decades, the number of tornadoes observed has increased markedly. For example, in 1967, 837 tornadoes were reported in the United States. This great increase over the average probably is a result of better reporting.

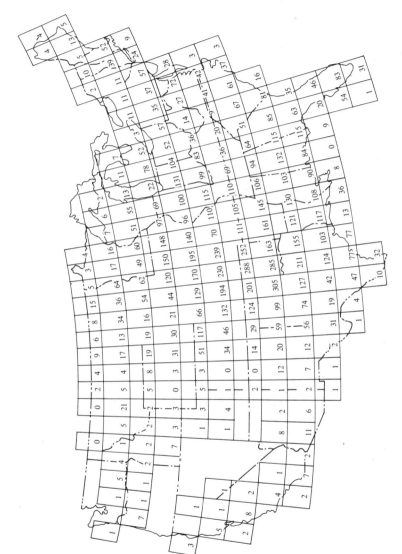

FIG. 6-16.   Total number of tornadoes by two-degree squares during a 45-year period (1916-1961). (Courtesy of ESSA.)

189

FIG. 6-17. Waterspout over Biscayne Bay, Miami, Florida. (Courtesy of G. E. Dunn, USWB.)

FIG. 6-18. Dust devil over Arizona. (Courtesy of P. C. Sinclair, University of Arizona.)

190

waterspouts. Usually, waterspouts are not as intense as tornadoes over land. Near their base, the winds churn the water surface, producing waves and spray.

A whirlwind that frequently forms on very hot days, especially over deserts, is the *dust devil* (Figure 6-18). Normally, there are no clouds associated with dust devils and they are no more than whirling columns of dust or sand. They are produced by strong convection near the surface, and given a rotation by slight, terrain-induced irregularities in the winds. These have been observed to rotate in both senses, clockwise and counterclockwise, with equal frequency.

PROBLEMS

Problems marked with an asterisk (*) are the most challenging.

*1. Compute the centrifugal force [Equation (4-6)] for the following circulation systems:

> Tornado: Radius = 0.5 km, speed = 250 km/hr.
> Hurricane: Radius = 50 km, speed = 150 km/hr.
> Extratropical cyclone: Radius = 500 km, speed = 50 km/hr.

Taking $2\omega \sin \phi$ equal to 0.25×10^{-4} sec^{-1} for the hurricane, and 1.0×10^{-4} sec^{-1} for the tornado and the extratropical cyclone, compare the magnitudes of the Coriolis and centrifugal forces in each case. Assuming a balance between the centrifugal force and the pressure gradient force in the case of the tornado, compute the pressure gradient if the air density is 1.2×10^{-3} g/cm^3.

2. Using the dimensions and speeds given in Problem 1, compute the kinetic energy per unit mass, $V^2/2$, for each vortex. Estimate the total volume of air circulating around each vortex, and from the mean density in the vertical (standard atmosphere table), calculate the total mass in each circulation system. Compute the total kinetic energy $(mv^2/2)$ in kilowatt-hours for each system. Compare the results with the electrical energy production in your city.

*3. Most schemes that have been suggested for modifying the weather can be easily shown to be impractical because of the enormous amounts of energy that they require. For example, it has been suggested that the smog of the Los Angeles Basin could be pumped into the interior desert. Suppose you wanted to change the air over the basin up to an altitude of 100 meters every hour over an area of 200 km^2. Neglecting any opposing forces due to gravity, pressure gradient, or friction, how much power would you need for this purpose?

4. List the energy sources of each of the major atmospheric vortices. Discuss the theories of how each vortex is initiated. Very few vortices are known to occur over polar regions in winter. Why?

5. During what season and in which hemisphere would you expect the sharpest transition zones (frontal boundaries) between air masses? Why?

6. In which sense does the vortex turn over your bathtub drain? Is the earth's rotation responsible? (Consider the magnitudes of the forces.) Would you expect any rotation if the bathtub were on the equator?

7

Climate

7-1. The Nature of Climate

A common misconception is to think of the climate of a region as the average state of the atmosphere. Not only do the average temperature, precipitation, wind, and other weather elements determine the climate, but also their variations. The diurnal, day-to-day, and seasonal changes, as well as the extremes of the weather, are all important in determining what crops can be grown, how homes and other buildings must be designed, and the way in which many other human activities may be conducted.

The major factors that control the climate are the same as those that produce the weather: (1) the intensity of solar radiation and its variation with latitude over the earth;* (2) the reflectivity (albedo) of the earth, which varies with the color, composition, and other characteristics of the earth's surface; (3) the distribution of land and sea; e.g., the preponderance of oceanic mass in the Southern Hemisphere compared to the Northern Hemisphere; (4) the effect of mountains and other topographic features on the atmosphere and its motions. In addition to large-scale influences, many local factors affect the small-scale or *microclimate:* vegetation characteristics, small bodies of water such as lakes, and even human activity that may alter the purity of the air and surface properties.

While a complete description of the climate would include a great many weather elements, the temperature and moisture of the atmosphere are the principal factors which control the broad-scale distribution of natural and cultivated vegetation, as well as the general nature of human activity which can be carried out in a region. Temperature and precipitation are also among the most easily measured of the atmospheric variables.

7-2. Temperature

The average, as well as the daily and seasonal variations of temperature are determined primarily by latitude, altitude, and the influence of land and sea. The influence of latitude on temperature can be seen in Table 7-1. It will be noted that while the *temperature* decreases from the equator to the poles (from low to high latitude) in both hemispheres, the *range of temperature* increases. The temperature decreases with latitude because of the greater incidence of solar radiation at the equator than at the poles, and the temperature range increases because both the incoming solar radiation and the outgoing terrestrial radiation have a much greater variation throughout the year at the poles than at the equator. The difference between the percentage of oceanic mass in the two hemispheres is reflected in the much smaller annual range in the Southern Hemisphere than in the Northern Hemisphere.

The latitude effect, as well as the difference between temperature regimes of locations close to the ocean and those in the interior of continents, is illustrated in Figure 7-1. Along the Atlantic coast, the temperature at Jacksonville, Florida, is generally higher at all times

---

* The word "climate" is derived from the Greek word "klimas," which means "angle of inclination." It referred to the angle of the sun above the horizon, which depends, of course, on latitude.

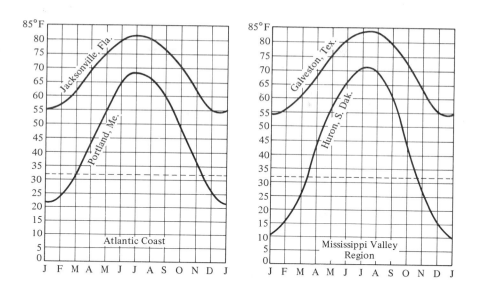

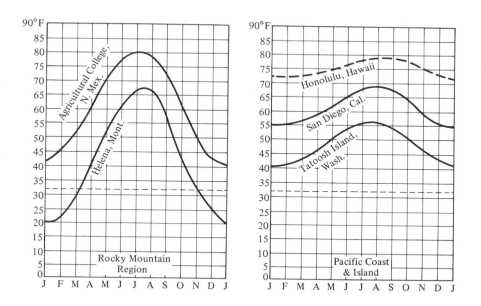

FIG. 7-1. Annual temperature variation at continental and marine stations.

TABLE 7-1

Mean Annual Temperature and Temperature Range, and Their
Variation with Latitude

Latitude (deg)	Mean temperature (°F)		Mean annual range (°F)	
	Northern Hemisphere	Southern Hemisphere	Northern Hemisphere	Southern Hemisphere
90–80	−8	−5	63	54
80–70	13	10	60	57
70–60	30	27	62	30
60–50	41	42	49	14
50–40	57	53	39	11
40–30	68	65	29	12
30–20	78	73	16	12
20–10	80	78	7	6
10–0	79	79	2	3

of the year than in Portland, Maine, but the range of temperature is greater at Portland than at Jacksonville. Similar latitudinal effects are present in the Mississippi Valley and in the Rocky Mountains. Along the west coast, the temperature at San Diego, California is higher than at Tatoosh Island, Washington; but because of the prevailing westerly flow, the temperature range at west coast locations shows little difference with latitude. A tropical island location like Honolulu, Hawaii shows a fairly even temperature throughout the year.

The range of temperature, both annual and diurnal, is an index of the *continentality* of a region; that is, the closer the location to the interior of a continent, the greater will be the average temperature range. The annual effect is illustrated by the temperature curves for Huron, South Dakota, a continental station, and Galveston, Texas, a coastal location. The diurnal effect is shown in Figure 7-2. El Paso, Texas, well within the interior of the continent, has a much greater temperature range between day and night than San Francisco, California, which is on the coast. The effect of altitude on temperature range is illustrated in Figure 7-3. In portions of the elevated southwestern United States, the mean difference between day and night temperatures in winter is over 30°F, while in the lower elevations of the Arizona and California deserts to the westward, the temperature range is about 20°F.

Temperature inversions. Although temperature typically decreases with elevation in the troposphere, certain special influences of topog-

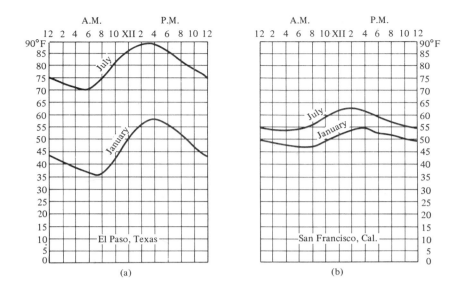

FIG. 7-2. Diurnal temperature variation at (a) a continental and (b) a maritime station.

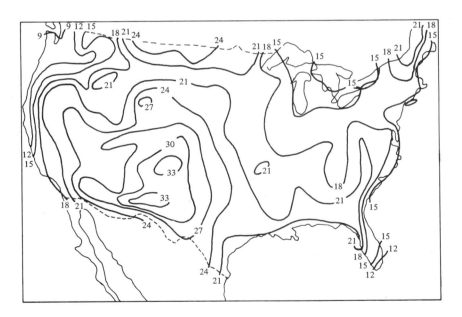

FIG. 7-3. Mean diurnal temperature range in January (°F).

raphy and atmospheric circulation may cause important deviations from the general rule. For example, temperature inversions (increase of temperature with height), caused by subsiding air at high levels in the atmosphere, are persistent features of the climate along the west coast of both South and North America. In populated areas, this temperature inversion traps the industrial and domestic smoke and fumes, producing a serious air pollution problem. On a somewhat smaller scale, Figure 7-4 shows the effect of elevation on the temperature during a clear, calm night. The temperature at the base of the slope in this agricultural location fell to well below freezing, while slightly over 200 feet higher, the thermometer remained above 50°F all night. This drainage of cold air toward low levels is well known in agricultural communities, so that farmers plant frost-sensitive trees and other crops along the slopes, leaving the bottom land for hardier plants.

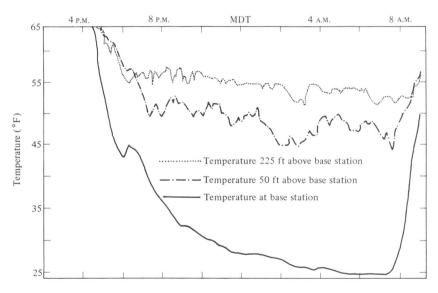

FIG. 7-4.   Continuous records of temperature from 4 P.M. to 9 A.M. at various heights on a steep hillside. (After F. D. Young.)

Because of its importance in planning the planting and harvesting of crops, frost (meaning, in this context, a temperature below 32°F), is a significant climatological measurement. In most parts of the world, the dates of frosts; i.e., freezing temperatures, which result in major damage to crops normally grown in the area, have been recorded for many years. Figure 7-5(a) shows the dates of the last killing frost in

(a)

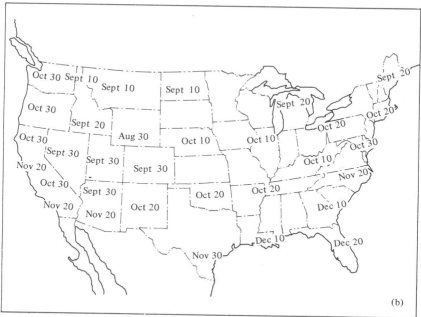

(b)

FIG. 7-5. (a) Average dates of last killing frost in spring; (b) Average dates of first killing frost in autumn.

199

spring in the United States, and Figure 7-5(b) shows the dates of the first killing frost in autumn.

Temperature indices. The distribution of temperature is, of course, an important factor in the comfort of the human body, itself. During the winter months, heating of homes and other buildings raises problems of supplying an adequate amount of gas, oil, or other fuel. To assist supply organizations in scheduling the delivery of heating fuel, an index called the *heating degree-day* is used. This index is based on the assumption that there will be little or no demand for heating at a given temperature, which is usually selected as an average temperature for the day of 65°F. The index is then defined as the number of degrees that the actual average temperature for the day is below the base temperature. (For example, if the average temperature on a given day is 50°F, the index would be 15.) By adding the daily values of the index over a period of time, heating engineers can determine the number of cubic feet of gas, or the number of barrels of oil, which will be required for heating purposes. For a typical city in the midwestern United States, Figure 7-6 shows total monthly degree-days

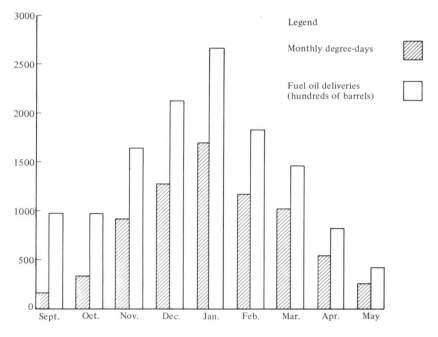

FIG. 7-6. Total monthly degree-days and fuel-oil deliveries in a midwestern city during an average winter. (After L. Crow.)

and fuel-oil deliveries during an average winter season. The month-to-month relationship between the two is evident. Figure 7-7 shows the annual average number of degree-days for the United States. The need for much larger heating facilities in the Rocky Mountains and the northern part of the country, compared with Florida and southern California shows up quite clearly in the chart.

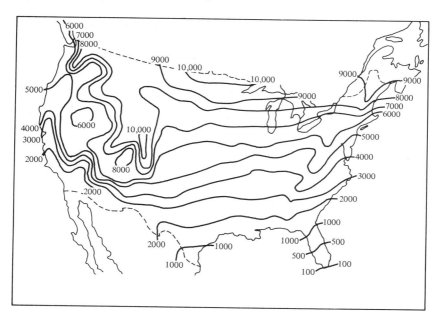

FIG. 7-7. Heating degree-days (base $= 65°F$).

A number of other indices of human comfort are used for special purposes. Similar to the heating degree-day is the *cooling degree-day*, which is defined as the number of degrees that the observed mean temperature for the day is above the base temperature (usually 70°F). (For example, if the average temperature on a given day is 80°F, the cooling degree-day index would be 10.) This index is used for estimating needs for air-conditioning equipment in homes and other buildings, and for scheduling electric power required to operate such equipment. Other indices have been devised by combining temperature with some other weather element; for example, the *comfort index* includes the effects of temperature and humidity, and the *wind-chill factor* is a combination of the effects of temperature and wind speed (Table 7-2).

TABLE 7-2

Wind-Chill Factor—An Index of the Cooling Power of the Wind. It is Expressed in Terms of the Equivalent Temperature without Wind.

Wind speed		Actual thermometer reading (°F)											
		50	40	30	20	10	0	−10	−20	−30	−40	−50	−60
Knots	mph	Equivalent temperature (°F)											
Calm		50	40	30	20	10	0	−10	−20	−30	−40	−50	−60
4	5	48	37	27	16	6	−5	−15	−26	−36	−47	−57	−68
9	10	40	28	16	4	−9	−21	−33	−46	−58	−70	−83	−95
13	15	36	22	9	−5	−18	−36	−45	−58	−72	−85	−99	−112
17	20	32	18	4	−10	−25	−39	−53	−67	−82	−96	−110	−124
22	25	30	16	0	−15	−29	−44	−59	−74	−88	−104	−118	−133
26	30	28	13	−2	−18	−33	−48	−63	−79	−94	−109	−125	−140
30	35	27	11	−4	−20	−35	−49	−67	−82	−98	−113	−129	−145
35	40	26	10	−6	−21	−37	−53	−69	−85	−100	−116	−132	−148

Wind speeds greater than 40 mph have little additional effect	LITTLE DANGER (For properly clothed person)	INCREASING DANGER	GREAT DANGER

7-3. Moisture

The supply of moisture used by man is dependent upon the processes of the hydrologic cycle—evaporation, condensation, precipitation, and runoff (see p. 45). Of these processes, all but runoff involve the atmosphere. However, a number of technological difficulties are present in the measurement of condensation and evaporation. The products of condensation, the clouds, cannot yet be measured quantitatively in any adequate manner; as we have already seen, our current observations of clouds are primarily descriptive in nature. Evaporation measurements are difficult to make and interpret because the evaporation depends upon the nature of the earth's surface, the type of vegetation, the temperature of the evaporating surface, the relative humidity of the air, and the wind speed. As a result of these complications, relatively few reliable evaporation measurements are available. Accordingly, most of the global information on the climate of moisture is based upon measurements of precipitation.

The global distribution of precipitation is linked strongly to the general circulation of the atmosphere and to the topographic features of the earth. The general circulation (Chapter 5) results in bands of upward vertical motion in the doldrums, where the trade winds of the two hemispheres tend to converge, and around the region of the polar front, where poleward-moving warm air rises over the cold air moving toward lower latitudes. Associated with these two belts of ascending air is found the greatest precipitation in both hemispheres (see Figure 7-8). In middle latitudes are located belts of descending air and a consequent decrease in precipitation; but the smallest amounts of precipitation are found in polar latitudes, where the air is also descending.

The combined influence of the atmospheric circulation and the topographic features of the earth can be illustrated by an examination of annual precipitation patterns from representative regions:

Equatorial. This belt extends about 10 degrees of latitude on each side of the equator; in it is located the "doldrums," or intertropical convergence zone. During the equinoxes, as the maximum of solar heating moves across the equator, the doldrums' zone of rising air moves with it. Thus, locations near the equator typically have two maxima of precipitation during the year, each occurring shortly after an equinox. Figure 7-9(a) shows the annual distribution of rainfall

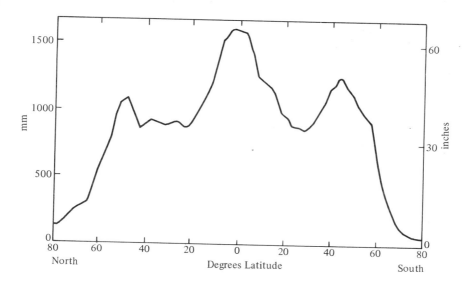

FIG. 7-8. Precipitation as a function of latitude.

at Yaounde, Cameroun (Lat. 4°N), which illustrates the two peaks in rainfall occurring in April and October.

Tropical. Beyond the range of the doldrums, but still in the tropical region, the effect of the equinoctial migration is not apparent. Here, the principal rainfall occurs during the summer season, and the winters usually have only small amounts of precipitation. Figure 7-9(b), showing the annual precipitation at Mexico City, Mexico (Lat. 19°N), is typical of the tropical region outside the belt of the doldrums.

Monsoon. Monsoon circulations occur along the coasts of most continents in lower middle latitudes, but they are most clearly defined along the coast of southern Asia. Monsoon precipitation occurs during the latter part of the summer season, and may be exceedingly heavy; the winter season is generally dry. Figure 7-9(c) shows the distribution of precipitation at Bombay, India (Lat. 19°N), where the monsoon is particularly well developed.

Subtropical. In the subtropics are located the "horse latitudes," which are characterized by a belt of semipermanent high-pressure cells. Around the eastern sides of the high-pressure cells are found the major deserts of the world; e.g., the Mojave in North America and the Sahara in North Africa. Rainfall here is exceedingly light, with some desert stations receiving no rain at all for periods of several years. Where rainfall does occur, as along the southwest coasts of Europe and North America, the precipitation is due primarily to intermittent passage of winter storms. Thus, the rainfall in these areas is characterized by dry summers and rainy winters, as is exemplified by Figure 7-9(d), which shows the annual distribution of precipitation at Sacramento, California (Lat. 39°N).

Middle latitude. In the region of the polar front, the warm, moist air being forced upward over the frontal surface is associated with the periodic middle-latitude storms. The amount and nature of precipitation occurrence is, however, greatly affected by a combination of the season and topography. Where the rain-bearing winds are forced over a mountain barrier, continental locations to the leeward receive little precipitation during the period of maximum storm activity; i.e., the winter season. This is exemplified by Figure 7-9(e), which shows the precipitation distribution at Omaha, Nebraska (Lat. 41°N). Here, the principal rainfall occurs from thundershowers during the summer, while the precipitation during the stormy winter season is quite light. Figure 7-9(f), however, shows that where the topographic influence is not present, as at Valencia, Eire (Lat. 52°N), the winter precipitation

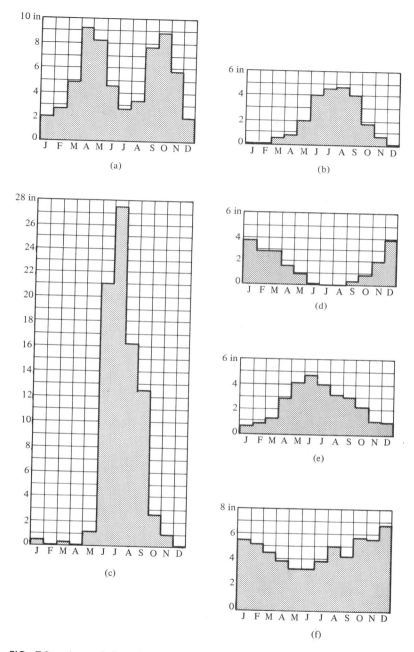

FIG. 7-9. Annual distribution of rainfall at: (a) Yaounde (Cameroun), (b) Mexico City (Mexico), (c) Bombay (India), (d) Sacramento (California), (e) Omaha (Nebraska), (f) Valencia (Eire).

205

is somewhat greater than during the summer, although, because of
its maritime location, significant amounts occur throughout the year.

Polar. In both hemispheres, the area surrounding the poles receives only a light amount of precipitation. For example, it is estimated
that the average annual precipitation over Antarctica is less than 17 cm
(7 in.) with a large portion having less than 5 cm. This is partly
because of the presence of a semipermanent high-pressure system over
the arctic and antarctic regions, and partly because the cold air contains little moisture. However, since almost all of the precipitation
falls in the form of snow and remains on the surface in solid form,
very little is lost by evaporation or runoff. Thus, even the small
annual amounts may continue to accumulate through the years, so
that massive fields of snow and ice may eventually build up in these
regions. The ice of the Arctic and Antarctic is removed primarily at
the edges where large chunks may fall into the sea to form icebergs.

Precipitation variability. Of all the weather elements, precipitation
is one of the most variable, both geographically and over periods of
time. In mountainous areas, locations on the windward side may
receive many times as much rain as falls on the lee side. The relatively
dry area on the lee side is called a *rain shadow;* examples can be seen
in Figure 7-10. Note, for instance, the arid region along the east
slopes of the Cascade Mountains and Sierra Nevada of the western
United States. Annual rainfall in the Olympic Peninsula along the
west coast of Washington frequently exceeds 100 inches, while in
the Wenatchee Valley 200 miles to the east, the yearly precipitation
is usually less than 20 inches. In the Southern Hemisphere, the arid
Patagonia area of Argentina and the desert "outback" of Australia
represent similar "rain-shadow" effects.*

Variability in rainfall over periods of time is also quite common.
The intermittent characteristics of precipitation may range from the
short-period variability of showers and thunderstorms, to such long-
period variations as "wet" years and "dry" years. The latter variations
represent particularly serious problems in agriculture, for most crops
depend upon a reasonably steady supply of moisture for proper
growth. In fact, a series of dry years in succession occasionally results
in such disastrous situations as the "dust bowl" years of the Great
Plains in the United States, and the severe drought conditions of
recent years in India.

_____

* The strong perturbations on the flow induced by mountain barriers is illustrated dramatically by the wave cloud of Figure 7-11.

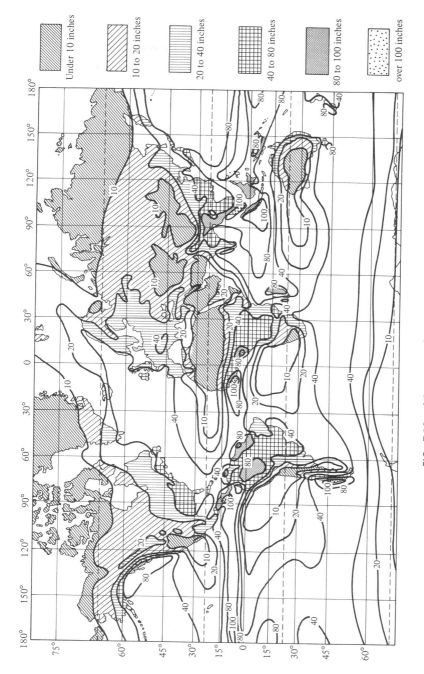

FIG. 7-10. Mean annual precipitation (inches).

Under 10 inches

10 to 20 inches

20 to 40 inches

40 to 80 inches

80 to 100 inches

over 100 inches

207

FIG. 7-11. A wave cloud over the Sierra Nevadas near Bishop, California. (Courtesy of H. Klieforth.)

The variability of rainfall may be measured by means of a "precipitation variability index," which is expressed by the following formula:

$$P_v = \frac{1}{N} \left(\frac{|R_1 - N| + |R_2 - N| + \ldots |R_n - N|}{n} \right)$$

where P_v = precipitation variability index
N = normal precipitation for the location
$R_1, R_2, \ldots R_n$ = annual rainfall for the location for each of n years (the symbol "|" means the difference is to be taken without regard to sign)

As an example, suppose the normal rainfall for a given location was 10.5 inches, and rainfall for five individual years was 12.2, 8.4, 9.1, 10.2, and 11.5 inches. The precipitation variability index would be:

$$P_v = \frac{1}{10.5} \left(\frac{|12.2 - 10.5| + |8.4 - 10.5| + |9.1 - 10.5| + |10.2 - 10.5| + |11.5 - 10.5|}{5} \right)$$

$$P_v = \frac{1}{10.5} \left(\frac{1.7 + 2.1 + 1.4 + 0.3 + 1.0}{5} \right) = \frac{1}{10.5} \left(\frac{6.5}{5} \right) = 0.124$$

A small value of the index indicates little variability in the precipitation, while a large value of the index indicates a high variability. Since it is desirable to plant crops in locations of dependable rainfall, other things being equal, those regions with low values of P_v generally provide good regions for agriculture. Some typical values for the United States are presented in Table 7-3.

TABLE 7-3

Values of the Precipitation Variability Index (P_v) for Representative Areas of the United States

Area	P_v
Northeastern U. S.	0.10–0.15
Ohio and Upper Mississippi Valley	0.10–0.15
Southeastern U. S.	0.15–0.20
Northwestern U. S.	0.15–0.20
Western Great Plains	0.20–0.25
Southwestern U. S.	0.30–0.40

In general, the variability of precipitation is low where the average amounts are relatively large, and the variability high where the average amounts of precipitation are small. Thus, the value of P_v for much of the British Isles is below 0.10, while in the Sahara Desert the value exceeds 0.40. A variability of 0.15 or less is typical of the two regions of rising air in the general circulation; i.e., the doldrums and the polar frontal zone. Periods of drought seem to be critical where the index ranges between about 0.20 and 0.25; where the variability exceeds these values, precipitation is usually so little that either no crops are grown or there is adequate irrigation to compensate for arid conditions.

7-4. Microclimates

The climate over large areas frequently shows small-scale variations due to the effects of minor topographic features, vegetation characteristics, and even such man-made structures as buildings, roads, and artificial lakes or reservoirs. These microclimatic variations are sometimes very important in determining the crops that can be grown and may affect many aspects of our health and comfort.

An example of the difference in temperature which may be experienced on a clear, calm night between a location in a valley and one on a hillside has already been illustrated in Figure 7-4. The effect that

TABLE 7-4

Microclimatic Difference between Windward and Leeward Shores of Lake Michigan (After Landsberg)

Climatic element	Time of year	Milwaukee (windward shore)	Grand Haven (leeward shore)
Mean temperature	January		3.6°F higher
Mean temperature	August	2.0°F higher	
Precipitation	Dec.–Feb.		2.09 in. more
Precipitation	June–Aug.	0.55 in. more	
Snowfall	Jan.–Dec.		44 days more
Sunshine	Dec.–Feb.	20% more	

Lake Michigan has on the microclimates in its vicinity is illustrated by Table 7-4. Grand Haven, Michigan, which is generally on the leeward side of the lake, has a higher temperature, more precipitation, more days with snow, and less sunshine during the winter than does Milwaukee, Wisconsin, which is on the windward side. During the summer, on the other hand, Grand Haven has a lower temperature and slightly less precipitation than does Milwaukee.

Trees and plants produce microclimatic variations primarily because of their effects on the moisture supply and the wind flow near the

TABLE 7-5

Qualitative Microclimatic Variations between Environments

Climatic element	Period	Forested areas	Open land
Mean temperature	Winter	Warmer	
Mean temperature	Summer		Warmer
Relative humidity	Entire year	Higher	
Wind speed	Entire year		Higher
		Valleys	*Hillsides*
Minimum temperature	Early morning		Higher
Maximum temperature	Daytime	Higher	
Fog	Night	More	
Frost	Early morning	More	
		Cities	*Suburbs*
Mean temperature	Entire year	Higher	
Wind speed	Entire year		Higher
Haze and "smog"	Entire year	More	
Precipitation	Entire year	Slightly more	

surface. Transpiration from vegetation causes locally higher humidity, while the soil tends to inhibit precipitation runoff. Temperatures and wind speeds are lower within forested areas than in the open. Cities tend to have higher temperatures than the surrounding suburbs (see, for example, Figure 7-12) and they produce more haze and smoke. Table 7-5 shows some generalized, qualitative differences between these microclimatic environments.

FIG. 7-12. "Heat island" over Corvallis, Oregon at 2200 PST, January 31, 1966. Note warmer areas over city and on hilltops in upper left and right center of photograph. (Courtesy of W. Lowry and Western Ways, Inc.)

7-5. Classification of Climates

The climate of a region is a composite of many different meteorological elements and their periodic variations, so that a complete description of the climate should take into account such parameters as the means, ranges, and diurnal and seasonal variations of temperature, precipitation, sunshine, wind, and other characteristics of the weather. However, not all of these elements are observed every-where over the earth. Furthermore, the nature of the climatic descrip-tion required may differ depending upon the purpose for which it is intended. For example, a description which was to be used to locate regions which would be desirable for planting of crops would be based upon different meteorological elements than one which was to be used for selecting an optimum site for an airfield. Accordingly, it has been found practical in designing a climatic classification system to limit the number of weather elements which need be considered, with the recognition that the system will have value for only certain limited uses.

Among the most widely known classifications of climates is the system proposed in 1918 by W. Köppen, a German climatologist. Its purpose is to relate the distribution and type of natural vegetation over the earth to those weather elements which most significantly in-fluence the growth of plants, and which, at the same time, are most universally measured. These elements are temperature and precip-itation. The Köppen climatic classification is therefore based upon the distribution of annual and monthly mean temperatures, and upon the seasonal variation in precipitation. Table 7-6 summarizes the cli-matic types and definition of each type in the Köppen system.

The primary classification is divided into six groups, each designated by a capital letter, ranging from "A" (tropical rainy climate) through "H" (undifferentiated highland climate). Subdivisions of the main groups are designated by adding additional letters representing further refinements in temperature and precipitation limits upon the primary groups. Figure 7-13 shows the world distribution of climate according to this system.

Another, more recent, climatic classification system proposed by C. W. Thornthwaite is based upon the presumed effectiveness of pre-cipitation upon plant growth. The "effective precipitation" is defined as the ratio of precipitation to evaporation at a given location. How-ever, because evaporation is difficult to measure and records are available at only a few places over the globe, Thornthwaite developed a formula which provides an approximate relationship between tem-perature and precipitation measurements and the precipitation/evap-oration ratio. In effect, therefore, the Thornthwaite classification, like the Köppen system, is based upon measurements of temperature and precipitation. Other classifications have been proposed to account for the "cooling power" of temperature, moisture, and wind upon the human body, but these have relatively limited application.

TABLE 7-6

Types of Climate

Groups of climate*	Types of climate	Precipitation†
A Tropical rainy (coldest month's temperature > 18°C)	Af, tropical wet	No dry season (driest month > 6 cm)
	Aw, tropical wet and dry	Winter dry season (driest month < 6 cm)
B Dry (evaporation exceeds precipitation)	BS, semiarid (steppe)	
	BSh, tropical and subtropical	Short moist season
	BSk, middle latitude	Meager rainfall, most in summer
	BW, arid (desert)	
	BWh, tropical and subtropical	Constantly dry
	BWk, middle latitude	Constantly dry
C Humid mesothermal (coldest month's temperature between 0° and 18°C)	Cs, dry summer subtropical	Summer drought, winter rain
	Ca, humid subtropical (warmest month > 22°C)	Rain in all seasons
	Cb, marine climate (warmest month < 22°C)	Rain in all seasons, accent on winter
	Cc, marine climate (warmest month < 22°C, less than 4 months > 10°C)	Rain in all seasons, accent on winter
D Humid microthermal (coldest month's temp. < 0°C; warmest month's temp. > 10°C)	Da, humid continental, warm summer (warmest month > 22°C)	Rain in all seasons accent on summer; winter snow cover
	Db, humid continental, cool summer (warmest month > 22°C)	Rain in all seasons accent on summer; long winter snow cover
	Dc, subarctic (less than 4 months > 10°C)	Meager precipitation throughout year
E Polar (warmest month's temp. < 10°C)	ET, tundra	Meager precipitation throughout year
	EF, ice cap	Meager precipitation throughout year
H Undifferentiated highlands		

With A climates:

    f = no dry season; driest month over 6 cm (2.4 in.)
    s = dry period at high sun or summer; rare in A climates
    w = dry period at low sun or winter; driest month under 6 cm (2.4 in.)

With C and D climates:

    f = no dry season; difference between rainiest and driest months less than in s and w; driest month of summer over 3 cm (1.2 in.)
    s = summer dry; at least 3 times as much rain in wettest month of winter as in driest month of summer; driest month less than 3 cm (1.2 in.)
    w = winter dry; at least 10 times as much rain in wettest month of summer as in driest month of winter

*† Precipitation and temperature values are averages.

213

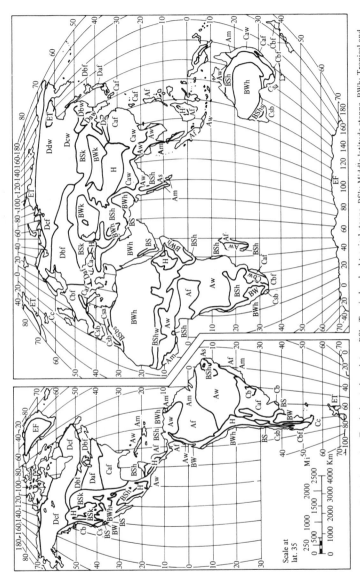

FIG. 7-13.   Climates of the world (Köppen).

Legend. Af, Am: Tropical wet. Aw: Tropical wet, dry winters. BSh: Tropical and subtropical steppe. BSk: Middle latitude steppe. BWh: Tropical and subtropical desert. BWk: Middle latitude desert. Cs: Dry summer subtropical. Cb, Cc: Marine. Ca: Humid subtropical. Cb: Humid subtropical. Da: Humid continental, warm summer. Db: Humid continental, cold summer. Dc, Dd: Subarctic. ET: Tundra. EF: Ice cap. H: Undifferentiated highlands.

214

Climatic extremes. Of considerable interest to almost everyone is "record-breaking" weather, and newspaper headlines periodically note the occurrence of a record low temperature or an excessively heavy snowfall at some location almost every year. Table 7-7 shows the highest

TABLE 7-7

Climatic Extremes. Highest and Lowest Observed Values of Certain Meteorological Elements

WORLD RECORDS	U.S.A. RECORDS
Temperature—Highest	
136°F at Azizia, Tripolitania, Sept. 13, 1922	134°F in Death Valley, Calif., June 10, 1913
Temperature—Lowest	
—127°F at Vostok, Antarctica, August 24, 1960	—76°F at Tanana, Alaska, January, 1886
Rainfall—Greatest annual	
1041.78 in. at Cherrapunji, India, August, 1860–July, 1861	578.00 in. at Puukukui, Maui, 1950 269.30 in. at Little Port Walter, Alaska, 1943 184.56 in. at Wynoochee Oxbow, Wash., 1931
Rainfall—Greatest monthly	
366.14 in. at Cherrapunji, India, July, 1861	107.00 in. at Puukukui, Maui, March, 1942 71.54 in. at Helen Mine, Calif., January, 1909
Rainfall—Greatest 24-hour	
73.62 in. at Cilaos, La Reunion, Philippines, March 15–16, 1952	38.20 in. at Thrall, Texas, Sept. 9–10, 1921
Longest period without rainfall	
19 years at Wadi Haifa, Sudan (entire period of record)	767 days at Bagdad, Calif., Oct. 3, 1912–Nov. 8, 1914
Snowfall—Greatest annual	
	1000 in. at Paradise Ranger Station, Washington, 1955–1956
Highest wind speed at surface	
231 mph at Mt. Washington, N.H., April 12, 1934	231 mph at Mt. Washington, N.H., April 12, 1934
Largest recorded hailstone	
	1.5 pounds at Potter, Neb., July 6, 1928

and lowest values of a number of weather elements, insofar as they represent extremes up to the current time. However, it is axiomatic that, by their very nature as extreme weather conditions, such records will continue to be broken. Probably by the time this text is published, at least one of these records will have been exceeded.

7-6. Climatic Change

Weather is noted for its variability. But how stable is climate? If one were to plot the value of some element of weather, such as temperature, as a function of time, fluctuations over all time scales would appear. For example, if the air temperature were averaged over hourly intervals, one would expect these averages to change from hour to hour in response to the earth's daily rotation. If these 24 hourly temperatures were then averaged for each day, one would notice day-to-day oscillations in these daily averages due to air-mass changes. Similarly, a distinct oscillation during the year is to be expected due to the earth's revolution around the sun. But if one were to average all the daily temperatures over each year, there would then appear year-to-year fluctuations, presumably due to changes in the general circulation of the atmosphere (Chapter 5). But even averages computed over decades and centuries fluctuate and, apparently, so do averages over thousands and millions of years, judging from the indirect evidence available. Thus, weather is not a constant on any scale of time, but rather appears to vary over all possible periods.

These cycles or periods (intervals of time over which the weather elements repeat themselves) are not, generally, very regular. Even the most regular of them—the daily and annual variations—are not unchanging. For example, the onset and ending of the summer monsoon over Southeast Asia vary considerably from year to year. In fact, forecasting from assumed periodicities in the weather has never been successful, precisely because of the great irregularity of such cycles, even for periods much less than a year.

Systematic weather observations do not exist for longer than about 300 years, so that long-period changes of climate must be inferred from evidence other than direct measurement. Most of such indirect evidence gives information only on precipitation and temperature. First of all, there are man's written records, which permit extrapolation backward in time for a few thousand years. Then there are the varying widths of growth rings of old trees, the migrations of people, the fluctuations in levels of lakes and rivers, and the succession of plant

TABLE 7-8

Climatic Variations During the Christian Era (After C.E.P. Brooks)

A.D.	Europe	Asia	Western North America	Africa
0	As present	Slightly rain-ier than now	As present	Good Nile floods
100	Somewhat drier	Rainy		Drier*
200	Rainy			
300		Dry		
400		Less dry Cas-pian −15 ft	Dry	Rainier
500	Drier		Dry	Rainy
		Dry		Rainy
600	Rather dry	Rainfall	Slightly rainier	
		increasing		Rainy
700	Dry, warm		Drier	Drier
		Rainy		
800			Dry period ended	Dry
	Rainier	Rainy in China		
900		Caspian	Rainier	Rainier
	Drier	+29 ft	Slightly drier	
1000				Drier
	Colder	Dry in China	Very rainy	
1100	Heavy rain	Dry Caspian −14 ft	Dry	Very dry
1200	Rainy	Dry		
	Very stormy	Rainfall increasing	Dry	Rainy
1300	Glacial ad-vance, drier	Rainy, Caspian, etc. high	Rainy	Rainy
1400				
	Glacial min.	Dry in China	Dry	Rainy
1500	Oceanic	Rainy, Cas-pian + 16 ft		Rainfall maximum
	Continental			
1600	Rapid advance of glaciers	Rainy Cas-pian + 15 ft	Rainier	Drier
1700	Dry in west Glacial max.	Near present Caspian rather high		
1800	Cold, rainier		Rainy	
1900	Rapid retreat of glaciers	Caspian falling	Drier	

* Comparative terms such as "rainier" and "drier" refer to change from previous period.

types which provide clues to the changes that have occurred during the past ten to twenty thousand years. The qualitative descriptions of the climates over some portions of the world during the past 2000 years, given in Table 7-8, have been deduced from such evidence.

Geological evidence must be used to extend the time scale further into the past. For this purpose, the types of flora and fauna, in fossilized form, found in an area are indicative of the climatic characteristics at the time they lived. The advance and retreat of glaciers (snow fields) which leave their imprint on the earth that they have traversed are signs of changing climate.

The estimated variation of temperature over the northern half of the Northern Hemisphere during the past 500 million years, deduced from fossils, is shown in Figure 7-14. There evidently have occurred great "ice ages" at intervals of about 250 million years: one some 700 to 1000 million years ago, another about 500 million years ago, a third 250 million years ago, and the latest, and best documented, which began roughly a million years ago and has only recently (geologically speaking) ended.

There have been numerous advances and retreats of the ice sheets not shown by the smooth curve of Figure 7-14. For example, in the last great Quaternary Ice Age, there have been at least four major advances and retreats of the ice. In the last of these advances, which

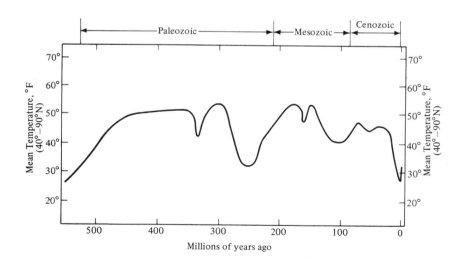

FIG. 7-14. Temperature changes in geologic time estimated from fossils. (After C. E. P. Brooks.)

lasted about 100,000 years, great ice packs extended hundreds of miles south of the Canadian border into the Mississippi Valley. Emergence from this last ice age began less than 20,000 years ago, with its completion only some 8000 years ago. Table 7-9 gives the climatic characteristics of Europe, according to Brooks, as that continent emerged from the last ice age.

TABLE 7-9

Climates of Europe Since Last Glacier (After C.E.P. Brooks)

Centuries ago	Climate	Vegetation
200–168	Arctic	Tundra
120–103	Sub-arctic/Temperate	Pine, sedge, peat
103–98	Arctic	Dryas°
98–88	Dry, cool	Pine, hazel
88–76	Dry, cold winters; warmer summers	Alder, oak, elm
76–45	Warm, humid	Peat, oak, alder, lime, elm
45–25	Drier, becoming cooler	Oak, giving place to pine
25–20	Cool, wet	Peat, beech

° Small, hardy, evergreen shrubs.

Beginning about 10,000 years ago, the European climate increased in warmth and dryness compared to the wetness and coldness of the preceding periods; between about 6000 and 3000 B.C., the climate reached an optimum—warm and humid—with annual mean temperatures about 2°C higher than today, and with year-round rain. The European climate cooled somewhat and was rather dry between 3000 B.C. and 500 B.C. It cooled markedly between 500 B.C. and 200 A.D.; during the period, the winters were snowy and cold and the summers cool and wet.

The evidence for climatic fluctuations over all scales of time is fairly conclusive. It has been shown in earlier chapters of this book that there are numerous factors that play roles in determining the state of the atmosphere: the energy received from the sun and its latitudinal distribution, the transparency of the atmosphere to solar and terrestrial radiation, mountain barriers, the frictional drag of the earth's surface. Just as there exists a variety of causes for daily, day-to-day, and week-to-week changes of weather, so too it appears that there must exist changes in one or more of these factors that were responsible for the longer-period, although irregular, changes of the earth's climates. The possible causes for these long-period climatic variations can be grouped into three broad categories: (1) *astronomical* (anything that alters the amount, type, or distribution of solar energy intercepted by the earth); (2) *atmospheric transparency* (anything that changes the transmission of radiation through the atmosphere); and (3) *earth's surface* (anything that alters the energy flow at the earth's surface or its geographic distribution). Some particular causes for long-period climatic changes within each group that have been postulated are listed in Table 7-10.

TABLE 7-10

Possible Causes for Climatic Change

Cause	Approximate range of periods induced (years)
I. Astronomical changes:	
A. Solar aging	10^9
B. Passage of solar system through galactic dust	10^8–10^9
C. Solar output variability	10^1–10^8 (?)
D. Earth orbit changes	10^4–10^5
II. Atmospheric transparency changes	
A. Volcanic dust in the stratosphere	10^0–10^8 (?)
B. Carbon dioxide content changes due to natural causes	10^4–10^8
C. Carbon dioxide content changes due to recent industrialization	10^1–10^2
D. Changes of other gaseous constituents	10^8–10^9
E. Dust particles introduced by man's activities	10^0–10^2
III. Earth's surface changes:	
A. Migration of the poles	10^7–10^9
B. Continental drift	10^7–10^9
C. Lifting of mountains and continents	10^7–10^9
D. Relative sizes of ice caps and oceans	10^4–10^5 (?)
E. Slow ocean circulation from great depths	10^3–10^6 (?)
F. Slow adjustments between atmosphere and ocean	10^0–10^3

It should be emphasized that no single factor is likely to be responsible for *all* of the observed changes in climate. All causes have had *some* effect; the problem is to determine which are most significant. Specific causes have been postulated to cover almost every observed period of fluctuation. We shall comment on only a few of the most seriously considered theories.

Astronomical. The most favored theory of this group is that the energy output of the sun, either in terms of quantity or spectral distribution, varies with time. Measurements over the past decades indicate that any variations in the *total* energy output are probably smaller than the accuracy of the measurements, which is about ±2 per cent. However, there are intermittent outbursts of corpuscles (charged particles) and very short wavelengths (ultraviolet) of radiation that are associated with disturbances on the sun that are largely absorbed in the outermost layers of the atmosphere (Chapter 1). These outbursts are known to affect the ionosphere and the earth's magnetic field and to produce auroras, but it is difficult to understand, at the present time, how these very short waves of radiation, absorbed by the very thin atmosphere above 30 or 40 km, can appreciably affect the great mass (more than 99 per cent) of the atmosphere that lies below, where the weather occurs.

These outbursts from the sun are prevalent during years of relatively high frequency of sunspots. Sunspots have been observed and counted since the days of Galileo. Well-defined oscillations in the number of spots appearing on the face of the sun occur over periods of about 11 and 23 years. Although periodicities of about the same lengths have been found for certain elements of weather, such as air circulation, pressure, and temperature, it is difficult to say whether such correspondence is merely accidental without knowing how the phenomena might be physically linked.

Just how changes in solar output might affect the climate is uncertain. It would appear unlikely that the effect would be merely to increase or decrease the average world temperature. Rather, an increase in solar output would result in greater heating in the equatorial regions than in the polar regions, thus increasing the latitudinal temperature gradient and therefore intensifying the atmospheric circulation. Stronger winds would lead, in turn, to increased evaporation, increased precipitation, and more extensive glaciation. Conversely, a decrease in solar output would reduce the temperature differences between poles and equator, thus weakening the atmospheric circulation, which in turn would lead to diminished glaciation. Very little is known about the variation in solar output during the earth's lifetime of about 4½ billion years. Until there is some direct evidence of abrupt changes in the sun's energy output, it is difficult to either accept or discount this theory as a significant cause of climatic change.

The second most favored theory in this group of "astronomical" causes is that of earth orbit changes. There are three types of variations in the earth's motions: (1) The obliquity of the ecliptic (i.e., the angle between the plane of the earth's orbit and the equatorial plane), which is now about 23½°, has changed by about 2½° in 45,000 years. When the angle is large, the seasons are more extreme and the pole-equator temperature difference in winter is increased. (2) Perihelion (point in the earth's orbit where the earth is closest to the sun) now occurs during early January, but the date of its occurrence advances through the year with a period of about 21,000 years. (Thus, in 10,000 years, perihelion will occur in July.) When winter in one hemisphere coincides with perihelion, it will be a little milder than when it coincides with aphelion. (3) The eccentricity (*ellipticity*) of the earth's orbit changes over a period of about 85,000 years. Although the difference in radiation received from the sun at perihelion compared to that received at aphelion is now only 7 per cent, with maximum eccentricity the difference is as much as 20 per cent.

Although these orbital changes only slightly affect the average yearly radiation received by the earth, by changing the seasonal distribution they may significantly alter the summertime melting and shrinkage of ice caps. Orbital changes may therefore be a significant producer of climatic change; however, they can hardly stand alone, since they cannot explain either the relatively short-period variations (less than 20,000 years) that have been observed or the long period without an ice age that occurred just before the last million years.

Atmospheric transparency. Volcanic eruptions sometimes spew fine particles of dust into the upper atmosphere that can persist for years (p. 9). These dust particles deplete some of the sun's rays by increasing the amount that is scattered by the atmosphere. In other words, increased dust causes an increase of the earth's reflectivity (albedo). For example, a mere 1-per cent increase in the earth's albedo could lower the earth's mean temperature about $1\frac{1}{2}°C$. Volcanic dust, trapped in the lower stratosphere, has been cited by some scientists as the cause of notable drops in the mean world temperature that have been observed in certain years during the past two centuries.

Dust counts at several places around the world indicate that the dust content of the atmosphere has been increasing steadily during the past few decades, perhaps due not only to volcanic eruptions but also to human activity. The mean world temperature has dropped by about $0.3°F$ in the past 25 years.

In addition, a variety of gases are given off by volcanoes, including carbon dioxide, which is a good absorber of infrared radiation in certain wavelengths (Figure 3-6). But there is some doubt as to whether an increased amount of carbon dioxide in the atmosphere would materially increase the absorption of infrared since the average concentration is already quite effective and, besides, its radiation absorption bands overlap those of water vapor. There is also some question as to whether the oceans regulate the concentration of carbon dioxide by absorbing any excess. Measurements of carbon dioxide content in the atmosphere are as yet too few to determine whether there are any long-term variations. There is evidence that the concentration of carbon dioxide in the air over cities is greater than in the countryside, and it may be that the increased consumption of fossil fuels, such as coal and oil, during the past couple of centuries has produced an increase of carbon dioxide in the air and, therefore, higher temperatures over urban areas.

Earth's surface. Anything that might alter either the properties of the earth's surface or their distribution over the globe could lead to climatic change. Evidence from magnetic data in rocks (*paleomagnetism*) has shown that there may exist a slow wandering of the positions of the poles on the earth's surface; i.e., in the past, the poles may have been at geographic points other than their present locations. The difficulty with this fact as an explanation of climatic change is that different areas would be affected at different times, but present evidence is that glaciers have occurred almost simultaneously over many regions of the earth.

Continental drift across the globe, possibly induced by convective currents within the earth's mantle, would certainly produce large changes in climate, but here again, they would not be simultaneous over many different land areas.

Mountain-building is generally considered to be an important cause of climatic changes, especially those taking several hundred thousand years or more. When a mountain is created, the increased elevation, in itself, gives rise to lower temperatures, but, in addition, it reduces the exchange between polar and equatorial regions. (For example, the Himalayas are quite effective in preventing the mixture of warm and cold air masses to their south and north.) But mountain-building and erosion take a long time and cannot explain the shorter periods of climatic change.

Expansion and contraction of the great polar ice caps would certainly have an important effect on world climates, principally because ice is such a good reflector (poor absorber) of solar radiation. Thus, an increase in the area of the ice caps would decrease the total amount of solar energy absorbed by the earth. It has been postulated, for example, that the great ice pack over Antarctica may periodically (about every 70,000 years?) slip into the adjoining oceans, spreading out to form continent-sized shelves. This slippage may occur when the ice depth over the Antarctic continent reaches some critical value such that the ice at the bottom, which is under enormous pressure, begins to melt. (Heat flowing from the earth's interior may help to melt this bottom ice.) If the ice were to extend itself over millions of square kilometers of ocean, the mean world temperature might fall several degrees centigrade, and sea level would rise some tens of meters.

The oceans represent a tremendous reservoir of heat. Recently, it has been shown that the cold water at great depths may slowly rise and mix with surface water. These deep water circulations may take tens of thousands of years to complete. Since the oceans cover such a large proportion of the earth's surface, they are, of course, an important source of heat in the atmosphere, so that changes in the ocean's surface temperature could result in significant changes of temperature and moisture in the atmosphere. However, it seems unlikely that such circulations could, by themselves, be responsible for the magnitude of temperature change observed during ice ages.

In summary, there are so many factors that control climate that it is little wonder that postulated causes for climatic change are so numerous. It would appear likely that more than one of the causes mentioned above is responsible for climatic changes, just as the daily weather is the result of many factors at work. But even a change in a single factor is likely to produce a complex reaction by the atmosphere. For example, an increase in radiation from the sun might cause this chain of events: The mean temperature is raised as well as the north-south temperature gradient; this may lead to increased air circulation and, therefore, greater convection and more cloudiness; the clouds might then cut off some of the radiation. Thus, the atmosphere may have a sort of "built-in thermostatic control."

Many of the theories that have been presented could produce some climatic change and perhaps did. The question that investigators try to answer is: Which is the most significant cause of the observed climatic changes? At the moment, two theories for climatic change are most popular: mountain-building and changing solar energy output. Until meteorologists can develop better models of atmospheric processes and of the general circulation, and geologists are able to more accurately date the climatic changes of the earth, the causes and prediction of the future climate can at best be highly speculative.

PROBLEMS

1. Why is there so much snow and ice over the Arctic and Antarctic even though precipitation is light? Assuming the average precipitation over

the Arctic is that of the average at latitude 80°N, what would be the minimum age of the ice at the bottom of a 100-ft iceberg? (Neglect compression of the snow.)

2. Why does the maximum precipitation occur in summer over the interiors of continents?

3. Classify the climate of your area of residence according to the Köppen system.

4. Explain the maxima of thunderstorms in Florida and over the Rockies (Figure 5-14).

5. From a knowledge of the general circulation, deduce the characteristics of the climate of the state of Washington, taking into account the topography.

6. All of the theories of climatic change involve the energy balance—changes in either the income or outgo. From what you know of the energy budget (Figure 3-9) and of atmospheric circulation, explain why it is unlikely that there exists a direct relationship between, say, increased solar output and climatic change.

7. Using the climatological data of Appendix 9, plot graphs of the annual temperature variation at several stations, selected to illustrate: (a) equatorial, (b) monsoon, (c) maritime, (d) polar, (e) continental climates.

8. Using the climatological data of Appendix 9, plot three graphs of the annual temperature range as a function of distance from the west coast for three latitudes in the United States. Start the first with Tatoosh Island and end it with Portland, Me.; the second should begin with San Francisco and end with Cape Hatteras; and the third should begin with San Diego and end with Jacksonville. When you plot each point, write the value of the annual mean temperature next to it. How much do the annual mean temperature and range vary with distance from the coastlines? Is there a difference between the west and east coasts? Why? Is there an effect of latitude? Of altitude?

8

Weather Forecasting

8-1. The Forecasting Problem

An important goal of all scientific endeavor is to make accurate predictions. The physicist or chemist who conducts an experiment in the laboratory does so in the hope that he may discover certain fundamental principles which can be used to predict the outcome of other experiments based on those principles. In fact, most of the so-called "laws" of science are merely very accurate predictions concerning the outcome of certain kinds of experiments. But few physical scientists are faced with more complex or challenging prediction problems than the meteorologist.

In the first place, the meteorological laboratory covers the entire globe, so that the problem of just measuring the present state of the atmosphere is tremendous. Furthermore, the surface of the earth is an irregular combination of land and water, each responding in a different way to the energy source, the sun. Then too, the atmosphere itself is a mixture of gaseous, liquid, and solid constituents, many of which affect the energy balance of the earth; one of them, water, is continually changing its state. Finally, the circulations of the atmosphere range in size from extremely large ones, which may persist for weeks or months, to minute whirls, with life spans of only a few seconds.

The problem of forecasting, then, involves an attempt to observe, analyze, and predict the many interrelationships between the solar energy source, the physical features of the earth, and the properties and motions of the atmosphere. Because of its complexity, weather forecasting forms one of the more advanced aspects of applied meteorology, and it cannot be treated in detail here. Nevertheless, our study of the weather would not be complete without examining this extremely fascinating branch of meteorology. We shall begin with a description of the manner in which the global weather service is operated.

8-2. How the Weather Service Works

The procedure used to carry out the process of observation, analysis, and prediction of the global weather requires cooperation among all the nations of the world. The activities of all nations are coordinated by the World Meteorological Organization, an international agency with headquarters in Geneva, Switzerland. Over one hundred countries are members of this body. The way the weather system works is illustrated schematically in Figure 8-1.

The first step involves *observations* of the current state of the atmosphere. More than 10,000 land stations and hundreds of ships make regular observations of the weather at the earth's surface, and more than 100 land and sea stations make radiosonde and rawinsonde measurements of the upper air (see Figure 8-2). In addition, weather reconnaissance aircraft take observations over oceanic areas where ship reports are scarce, and weather satellites provide data on clouds and radiation.

Most of these observations are made synoptically (a term which, in meteorology, means not only "in an overall summation," but also

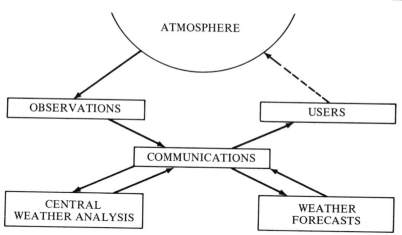

FIG. 8-1.   How the weather service works.

"simultaneously"). By international agreement, the regular synoptic observations are made every six hours beginning at midnight Greenwich Mean Time (0000 GMT, 0600 GMT, 1200 GMT, and 1800 GMT) each day. In addition, observations for aviation purposes are made at many airports every hour, or more often if the weather is changing rapidly. Specialized observations may also be made under certain conditions for agricultural, industrial, research, or other purposes. These observations are then fed into a communications subsystem, usually in a standardized message format. Because the accuracy of weather forecasts decreases rapidly with time, a primary goal of meteorological communications is speed. Thus, most weather reports are transmitted by electronic methods: radio, telephone, teletype, and even photo-facsimile.

The observations are collected at a number of points where they are processed by a central weather analysis organization. Three locations have been designated by the World Meteorological Organization as World Meteorological Centers; these are located in Melbourne, Australia, Moscow, U.S.S.R., and Washington, D.C., U.S.A. In addition, most countries maintain national centers, where the basic weather needs of the domestic economy are met. In the United States, the National Meteorological Center is also located in Washington, D.C.

The products of these central analysis groups are basic "guidance materials," primarily for the use of meteorologists. They consist largely of maps and computations that describe the current state of the weather: the locations of cyclonic and anticyclonic circulations, frontal

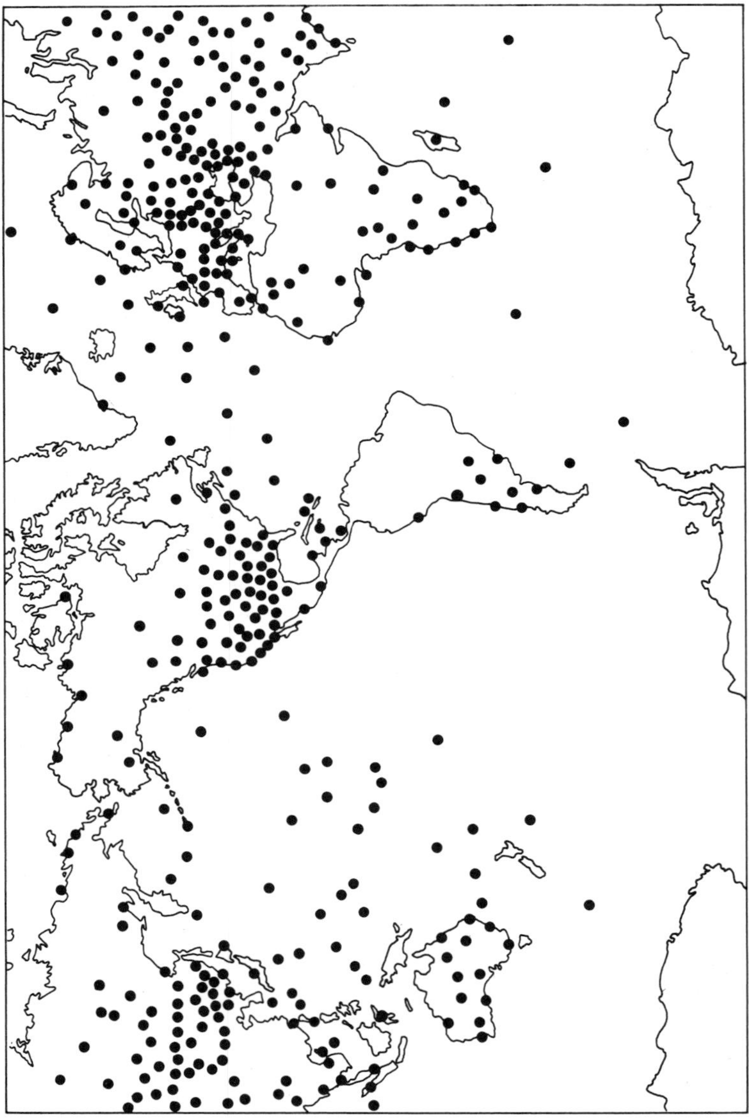

FIG. 8-2. World radiosonde network.

types and positions, regions of atmospheric instability, and many other characteristics of the atmosphere on a large scale. In addition, maps and charts of predicted conditions are prepared.

The guidance material is transmitted by the weather analysis centers to cities and towns throughout each country. Typically, a local meteorologist prepares the weather forecast, interpreting the technical information that he has available both from the analysis center and in his immediate vicinity, in terms of rain, snow, ceiling, visibility, or other data required for local needs.

These weather forecasts are transmitted by radio, television, newspapers, and telephone to the ultimate users of the information. Of course, for some purposes, the final forecast may be made by a meteorologist specializing in aviation operations, agricultural problems, industrial activities, and the like. In this case, the prediction is interpreted in terms most useful for, and is provided directly to, these users.

In some cases, the user of the forecast may find it possible to modify certain aspects of the weather which are undesirable. For example, a farmer may operate orchard heaters in order to protect against predicted frost, or an airline may seed low clouds which are interfering with landings. This modification of the atmosphere is indicated in Figure 8-1 by a dotted line. Although, at present, such modifications are only possible in relatively small areas, as man's technology improves, large-scale modification will undoubtedly be attempted. (See Chapter 10.) Then the meteorological system will be truly a "closed circuit," with more frequent observations of the atmosphere being required to keep up with the atmospheric changes which man himself has caused.

Weather-Map Analysis

After the weather observations have been made and transmitted to a weather analysis office, the first step in making the weather forecast begins. This is the preparation of a pictorial three-dimensional description of the atmosphere. While many techniques are used for carrying out this procedure, the most widely used involves the construction of a series of charts that represent horizontal "slices" of the atmosphere. These charts are prepared for predetermined levels extending from the earth's surface to well into the stratosphere.

The surface chart is by far the most complete, both in terms of the number of stations reporting and the number of atmospheric properties measured at each point. This is the familiar weather map that appears, usually in somewhat simplified form, in many newspapers. Examples of hemispheric surface weather maps are given in Figure 8-3.

The construction of the surface chart begins by plotting, at the geographic location of each station, the primary weather elements observed at the station. The position of each element is specified by a "plotting model," which is given in detail in Appendix 5. As will be noted from the model, a great deal of information is depicted by this device: wind direction and speed; pressure; pressure change in the past three hours; temperature; dew point; visibility; current weather (snow, rain, etc.); weather in the past six hours; cloud type, amount, and height; and precipitation type and amount.

Following the plotting of these data over the entire map, the weather analyst constructs isobars (lines of equal pressure) so that he can determine the position, size, and intensity of high- and low-pressure centers. During this process, the locations of frontal systems are determined. This is done by considering, first of all, the position of fronts on the previous weather chart, and the probable movement and development of entire frontal systems during the intervening period. A basic consideration in this process is the polar front model discussed in Chapter 6.

In principle, the process of surface weather-map analysis is illustrated in Figure 8-4. In this schematic chart, the plotting model has been simplified to show only the most significant (for this purpose) weather elements; i.e., wind direction and speed, pressure, pressure change in the past 3 hours, temperature, and current weather. The position of the frontal system twelve hours earlier is indicated by the dashed line.

Considering first the analysis of the pressure field, it will be noted that each isobar represents the analytic estimate of the location of the pressure indicated by its labeled value. For example, the isobar labeled 1008 (top left-hand side of the map) passes between one station where the pressure is 1009.2 mb (092, with the first two figures omitted), and an adjacent station where the pressure is 1007.0 mb (070). Clearly, the isobar with a pressure of 1008 mb must lie between the two stations. Following this isobar over the entire map, it will be observed that this principle is followed throughout its length. Other isobars are constructed in a similar fashion.

Now, taking a look at the frontal system, it will be seen that all fronts have moved generally eastward (to the right), but the cold front has tended to move faster than the warm front. (Normally, it would eventually catch up and form an occlusion.) The precise location of the fronts can be confirmed by referring to the weather data plotted

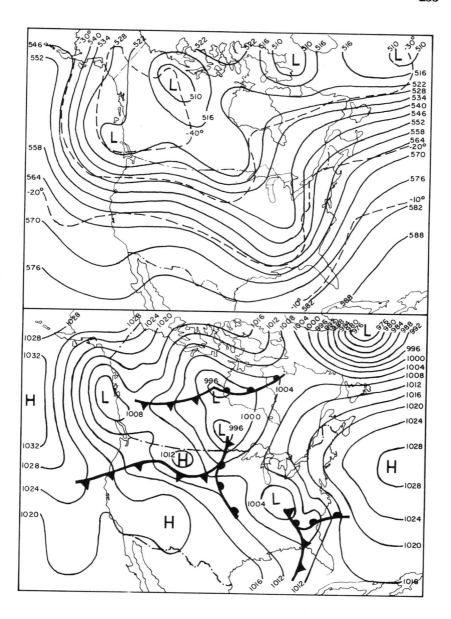

Fig. 8–3. A sequence of sea level and 500-mb charts for North America. (See Appendix 5 for meaning of symbols.) (a) March 1, 1966; 1800 GMT.

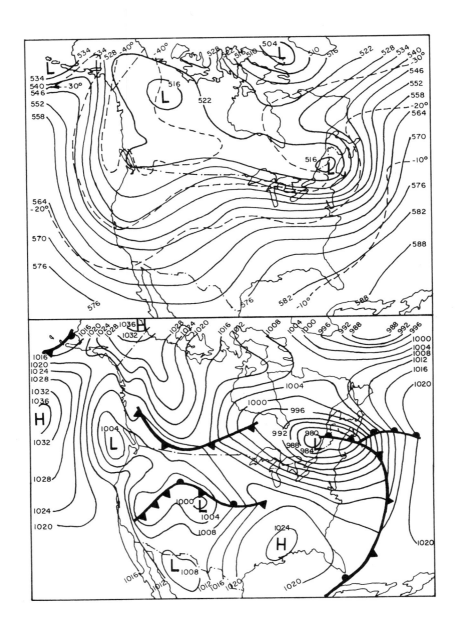

(b) March 2, 1966; 1800 GMT

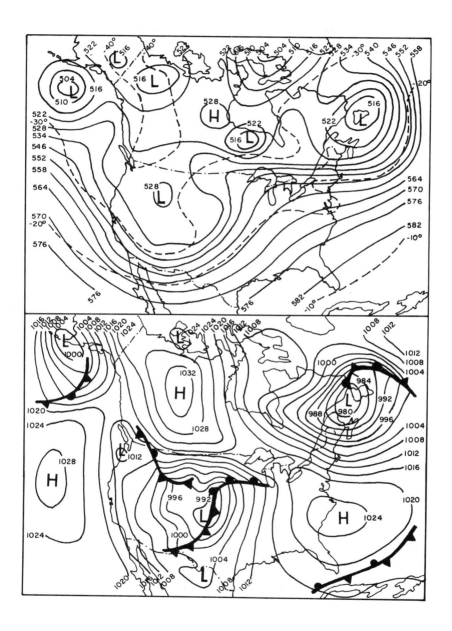

(c) March 3, 1966; 1800 GMT

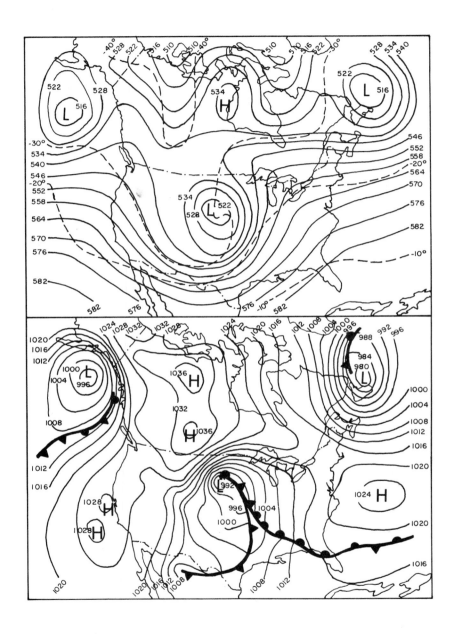

(d) March 4, 1966; 1800 GMT

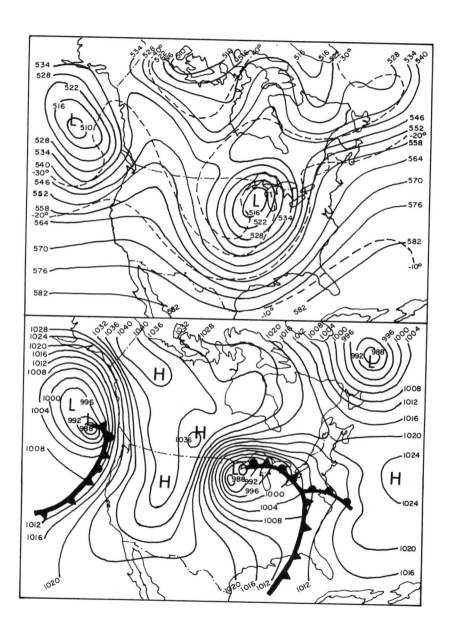

(e) March 5, 1966; 1800 GMT

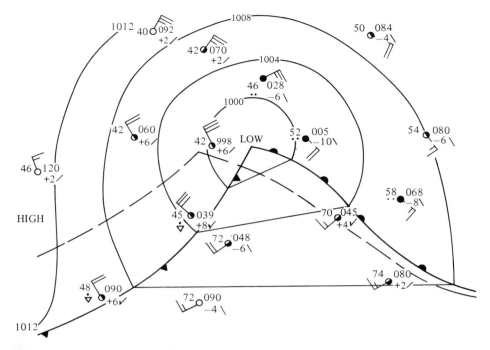

FIG. 8-4.   Example of a sea level weather chart analysis.

beside each station. For example, note that ahead of the warm front, temperatures are generally in the cool 50's, winds are mostly south-easterly, the barometric tendency in the past 3 hours has been down-ward, and there is steady rain near the front. In the warm sector, between the cold and warm fronts, temperatures are in the 70's, and winds are southwesterly. Behind the warm front, barometric tendencies show a slight rise, and the steady rain has ceased. Ahead of the cold front, pressures are beginning to fall, and right at the front there are showers. Well behind the cold front, temperatures are in the 40's, the wind is from the north, pressures are rising steadily, and there is little or no precipitation.

Of course, in this simplified model, many of the complexities of the real atmosphere have been omitted. The process of occlusion, for example, will introduce many departures from this ideal situation, and the presence of old, decaying fronts, or developing new ones, will add to the difficulties of analysis. The surface maps of Figure 8-3 illustrate some of the more complex analyses encountered in actual practice.

Since the weather at the surface depends to a considerable degree on the properties of the atmosphere at higher levels, the meteorologist must confirm his analysis by a study of the temperature, moisture, and wind flow aloft. Conditions in the upper air are usually represented by charts of isobaric (constant pressure) surfaces. These are equivalent to charts of constant altitude, since the horizontal pressure gradient at a constant level is proportional to the slope of an isobaric surface (see Figure 4-2). Furthermore, the geographic relationship of low to high pressure on a constant altitude surface is directly equivalent to the relationship of low to high heights on a surface of constant pressure. This is illustrated in the schematic cross section of the atmosphere shown in Figure 8-5.

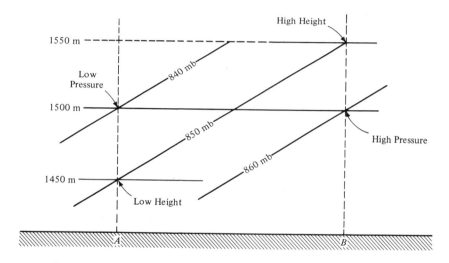

FIG. 8-5. Relationship between constant altitude surfaces and constant pressure surfaces.

The line designated 1500 m represents a level surface which is intersected by the pressure surfaces labeled 840 mb, 850 mb, and 860 mb. Above point *A* at 1500 m the pressure is 840 mb, but above point *B* at the same level the pressure is 860 mb; *at 1500 m, therefore, the pressure is lower at A than at B.* Above point *A* at 850 mb the height is 1450 m, but above point *B* at the same pressure the height is 1550 m; *at a pressure of 850 mb, therefore, the height is also lower at A than at B.* Accordingly, on constant pressure charts, not only does the wind conform to the gradient of contours in the same way that the wind is determined by the isobars on a constant altitude chart, but the relative positions of "lows" and "highs" are the same on both charts.

The use of constant pressure, rather than constant altitude charts aloft is primarily for convenience. Computations of the pressure gradient force at different altitudes in the atmosphere require that the change in density with height be taken into account. However, if constant pressure charts are used, so that height gradients are substituted for pressure gradients, the effect of variations in density is automatically accounted for. Thus, for example, computations of the geostrophic wind require adjustments for the density on different constant level surfaces, but not on constant pressure surfaces.

Examples of global analyses of 500-mb charts (approximately 5500 m) are shown in Figure 8-3. These upper air maps are constructed using the same principles as those described for surface charts. Meteorological data obtained from radiosonde or rawinsonde observations are plotted at the geographic location of each station, using the plotting model shown in Appendix 5. After these data have been plotted over the entire chart, contours representing the height of the selected pressure surface are constructed. The constant pressure surfaces that are routinely analyzed are 850, 700, 500, 300, 200, and 100 mb (approximate altitudes are, respectively, 1500, 3000, 5500, 9200, 11,800 and 16,200 m).

A simplified example illustrating this type of analysis at 500 mb is given in Figure 8-6. The plotting model is abbreviated to show only the wind direction and speed, temperature, and height of the 500-mb surface at each station. The analysis of contours ("topography" of the pressure surface) is carried out in a manner similar to that of isobars on a constant altitude chart. For example, the contour labeled 5400 m starts at the left-hand side of the chart and passes between one station where the 500-mb height is 5460 m (546, with the last unit omitted), and an adjacent station where the height is 5360 m (536). The height of 5400 m therefore lies between the two stations, and the remainder of the analysis is constructed in a similar manner.

On upper-level charts at this elevation and higher, fronts are rarely indicated, since many do not reach these heights and, of those that do, upper-air observations are frequently too sparse to show the locations of frontal boundaries. Note, however, that temperatures on the downwind (right-hand) side of the LOW are relatively high, signifying the overrunning of tropical air above the warm front at lower levels. Upwind of the LOW, the lower temperatures are associated with the arctic or polar air behind the cold front. Because surface friction is absent, winds at this level are nearly geostrophic (parallel

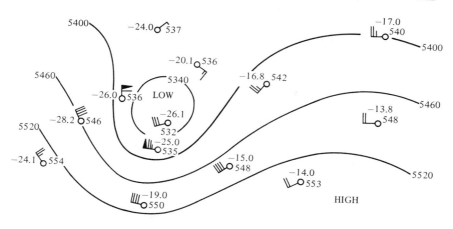

FIG. 8-6. Example of a 500-mb chart analysis.

to the isobars), and are strongest just south of the main LOW center, where the jet stream would normally be found.

Many other types of analyses are regularly made at most meteorological forecast centers: vertical cross sections of the atmosphere showing the distribution of temperatures, winds, and moisture in areas of particular concern; analyses of wind speed that help to locate the jet stream; analyses of atmospheric stability for the prediction of thunderstorms and tornadoes; and others that describe the properties of motion (e.g., the air's "spin"), potential energy, etc.

Meteorological Prognosis

After the current state of the atmosphere has been determined by analyses such as those described in the previous section, the next step is to make a meteorological prognosis. For relatively short periods (up to 48 hours or so), this is done by projecting the movement of high- and low-pressure systems, and their associated fronts, into the future. At the same time, the possible development or decay of these attributes of the atmosphere must be determined. Some idea of the difficulty of the problem may be obtained by an examination of the surface and 500-mb charts of Figure 8-3.

In this figure, notice that the sea level chart is generally dominated by several closed isobaric systems—cyclones and anticyclones—but, aloft, the picture changes considerably. There are few closed systems; rather, a general westerly flow undulates around the poles in middle

latitudes, where the westerlies usually attain peak speeds. Four or
five major waves in the westerly flow pattern normally can be identified
around the hemisphere, with many minor waves superimposed. These
small waves are associated with fast-moving, short-lived (3 or 4 days)
wave cyclones at the surface. The longer waves, however, are related
to the larger-scale, more sluggish features of the weather. When a
particular region of the globe is located to the east of the crest and
west of the valley of one of these long waves, there is likely to be a
2- or 3-week period of dry weather. On the other hand, if it is to the
east of the valley, wet weather is likely to prevail for a few weeks.

From this generalized picture of the hemispheric weather, a
more detailed examination must be made of the smaller-scale features.
For example, consider Figure 8-7, where a simple frontal analysis has
been superimposed on a map of the midwestern United States. We
see that the cold front has passed Amarillo, Texas, and an examination
of its position 12 hours before shows it to be traveling toward the

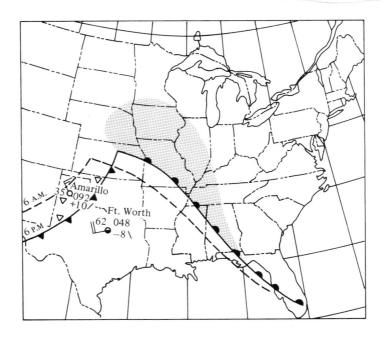

FIG. 8-7. Simplified frontal analysis and associated weather. (Solid
lines represent current frontal positions; dashed lines depict
positions of fronts 12 hours earlier.)

southeast at about 15 mph; this is confirmed by the wind at Amarillo (from the northwest at 15 knots). Since the front caused showers as it passed Amarillo, and it is about 180 miles from Fort Worth at 6 P.M., a preliminary estimate of the forecast would be "partly cloudy tonight, turning colder with showers around 6 A.M. tomorrow."

This is, of course, a very simple method of forecasting. It might be called a "steady-state" method since we have assumed that the past movement of the front and its associated weather will continue at a steady rate for a short time into the future. However, this assumption may not always be realized. The front may slow down or speed up, dry air may replace the moisture as the front moves over Texas, absence of solar heating during the night might result in less lifting at the front, and many other influences may affect the weather by the time the front reaches Fort Worth. In part, some of these additional influences may be accounted for by the forecaster's experience in a particular region.

While the above procedure will usually produce reasonably accurate forecasts for periods of a few hours, other methods must be applied for predictions of a day or two. One of these involves the construction of a "prognostic chart," where some attempt is made to account for such influences as the intensification or weakening of the circulation around a low center and the acceleration of fronts. The principle is illustrated in Figure 8-8.

Comparing Figure 8-8 with Figure 8-7, it will be noted that the cold front speeded up between Monday evening and Tuesday morning, at which time it was moving toward the southeast at a speed of about 40 mph (note the wind direction and speed at Fort Worth). This means that the cold front was "catching up" with the warm front near the apex of the frontal wave, and the process of occlusion had begun. A prognosis of the frontal pattern during the following twelve hours should take account of this acceleration, but the speed of the front may not increase at a steady rate. Since the rapid movement of the cold front will lead to rapid occlusion, and the occlusion process gradually decreases potential energy of the system (Chapter 6), the circulation intensity will tend to decrease. Thus, the frontal acceleration will not be constant, but rather will decrease with time. An estimate of the frontal position twelve hours later (6 P.M. Tuesday), based on the decreasing rate of acceleration, is shown in Figure 8-8. The typical weather associated with this prognostic chart is also shown; at 6 P.M. on Tuesday the likely weather at Pittsburgh, for example, would be "cloudy with steady rain."

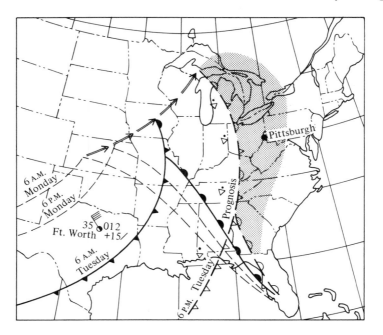

FIG. 8-8. Schematic prognostic chart showing acceleration of frontal system. *using storm tracks See next page*

While we have considered in these examples only the predictions of the fronts and their associated weather, the movement of the low-pressure center, since it is normally the region of greatest storm intensity, is also of considerable importance. The projected track of the low-pressure center (shown by the dashed arrow in Figure 8-8) can be estimated by a number of methods. Perhaps the simplest is to project its motion along the path normally taken by low-pressure centers in the region in which the storm originates. Such paths for the winter season in the United States are shown in Figure 8-9. Average storm tracks for other seasons are generally similar, although there are, of course, fewer storms during the warmer part of the year. While the forecast accuracy deteriorates considerably when the projection is made for long periods into the future, this procedure provides a good first approximation to the storm movement.

However, if one compares the normal storm paths of Figure 8-9 with the average flow pattern in the upper air shown in Figure 5-3, it is evident that the low-level storm centers move in the same general direction as the upper-level flow. This suggests that there is an inter-

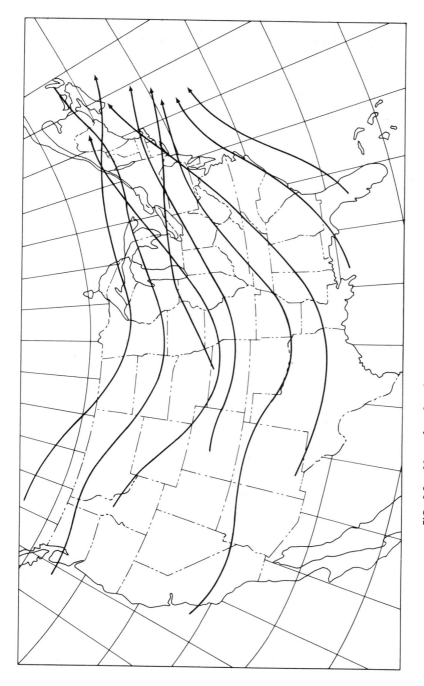

FIG. 8-9. Normal paths of winter storms. (After Bowie and Weightman.)

connection between the movement of the storm (low-pressure) centers at the surface and the high-level winds; i.e., the storms move with the upper winds. Furthermore, because the upper-level flow pattern is relatively stable and changes only slowly with time, it is possible to use this relationship to refine the projected motions of surface storms.

Figure 8-10 shows an occluding storm which originated in western Kansas. Its normal path, as indicated by Figure 8-9, would carry the storm center due eastward through Missouri, and then east-northeast-ward over Illinois, Indiana, Ohio, and into New York State. However, this trajectory could be greatly modified by the upper-level flow in which the storm center is imbedded. In Figure 8-11, the 500-mb contour pattern shows cyclonic flow around a deep center located over eastern South Dakota. Thus, if the upper-level pattern were to remain station-ary, or move only slowly, the path of the surface storm center would move as indicated by the dashed arrows eastward through Missouri, then northward through Illinois, Wisconsin, and over Lake Superior.

Such "steering" of surface low-pressure centers by the flow at higher levels affects the movement of many kinds of storms. Hurricanes,

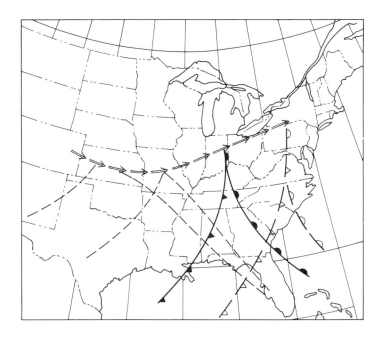

FIG. 8-10. Expected path of Kansas storm, if it follows the normal trajectory (Figure 8-9).

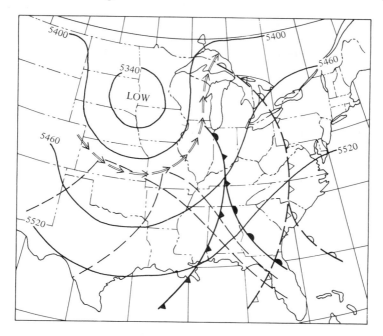

FIG. 8-11. Expected path of Kansas storm, if it follows the 500-mb flow.

for example, are known to move around the periphery of high-level anticyclones and then may be caught up in the westerly flow aloft as they reach higher latitudes (Chapter 6).

Prognosis by electronic computers. Meteorologists have long dreamed of being able to predict the future state of the atmosphere by equations, much as the astronomer determines the occurrence of a future eclipse. But while the basic physical equations governing the behavior of fluids in general have been known for almost a century, and were developed for the atmosphere more than fifty years ago, only within the last fifteen years has it become possible to apply these principles in practice. The development which made this feasible was the invention of the electronic computer.

The basic physical equations that must be solved by the computer are those that relate the change with time of each atmospheric variable quantity (e.g., wind velocity, pressure, density) with some combination of the other variable quantities. For example, Equation (4-1), which is the meteorological form of Newton's second law, relates the rate of change of the velocity of a parcel of air with the distribution of pressure and density at the position of the parcel. The acceleration of each parcel of air can be computed from Equation (4-1), if the direction and magnitude of each of the forces acting on the parcel can be specified. Of course, at the very beginning of a forecast period, the forces can be measured everywhere from the observations of pressure and density (or temperature).

Starting with an initial, observed, field of net force, the future velocity and location of each parcel of air can be computed from Equation (4-1) by using the procedure illustrated in Figure 8-12, if it is assumed that the force field does not change over the time period of the forecast.

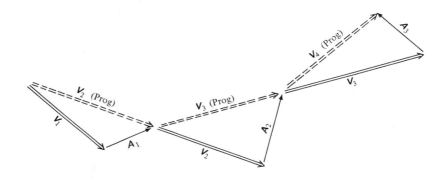

FIG. 8-12. Principle of stepwise prognosis used in numerical weather prediction.

The vector V_1 in Figure 8-12 is the observed initial velocity of the air at a particular location; the length and direction of the solid double arrow therefore represents the path of an air parcel traveling at the velocity V_1. The vector A_1 is the acceleration experienced by the parcel as a result of the forces acting on it—the pressure gradient force, Coriolis force, etc., as given by Equation (4-1). Thus, if these forces are measured at the initial time from a weather map, their total effect on the velocity of the parcel (the net acceleration) can be computed from this equation. This value, A_1, when multiplied by the time increment and added vectorially to V_1, yields a prognosis of the velocity of the parcel, V_2 (Prog), at the end of the time period.

At the end of this time, the parcel will have traveled to the point indicated by the head of the dotted arrow. Its new velocity is shown by the solid arrow labeled V_2. The net force field will be changing continuously as the patterns of pressure and density change. Therefore, at the end of the time interval between the first and second positions of the air parcel, the complete set of equations must be solved simultaneously for all air parcels to determine the new force field. When this has been done, the new acceleration, $\mathbf{A}_2$, will be known and can be added to the velocity, $\mathbf{V}_2$, to provide a new estimate of the velocity for an additional short time in the future. This is shown by the dotted arrow, $\mathbf{V}_3$ (Prog). The entire process can be repeated, producing a further prognosis, $\mathbf{V}_4$ (Prog), and so on.

When applied to meteorological prognosis, this stepwise procedure is frequently known as "numerical weather prediction." In principle, the process of computing the prognostic velocity vectors for each time step could involve many billions of parcels of air in the atmosphere, depending on the size of the parcels into which we subdivide the atmosphere. In practice, however, the procedure must be simplified by using a grid system which takes account of "parcels" which are several hundred miles apart. The length of the time step varies, but may be as short as 10 minutes. This means that a tremendous amount of calculation must be performed in order to produce a prognosis for even short periods of a day or two. For example, for a 24-hour prediction of the North American flow, more than half a billion arithmetical operations must be carried out. Obviously, such computations cannot be performed in a reasonable time by human effort alone—the immense power and speed of an electronic computer is required. And even with such devices, many simplifying assumptions must be made in the equations before they can be solved in a practical way.

In addition to the sheer magnitude of the computational task, there are other difficulties in making accurate predictions. One of these is that observations exist in insufficient detail over much of the earth to permit an accurate representation of the fluid motion everywhere. Meteorologists have had to confine their numerical solutions to areas where there is a reasonable density of atmospheric measurements. But the weather 1000 miles away may provide the "seed" for tomorrow's weather in the forecast area; i.e., no portion of the atmosphere is forever completely independent of the rest. Thus, the effects of the areas excluded from the prognostic equations gradually introduce errors that increase with time.

Another problem arises because the *net* force acting on any part of the atmosphere is generally very small compared to each of the *individual* forces which are present. For example, in the vertical, the force of gravity acting downward on one gram of air and the vertical pressure gradient force acting upward are each individually very nearly 980 dynes. In a given situation, the difference between these two, or the net force, may be only 1 or 2 dynes. This means that to compute the net force in the vertical, one must measure the vertical pressure gradient everywhere with a very high degree of accuracy. In most cases, the accuracy required is greater than can be achieved with the present techniques and density of measurement.

Finally, there are some technical difficulties of how to take into account the highly variable effects of friction, mountain barriers, and the distribution of heat and cold sources. But, despite the many problems and the need to make simplifying assumptions in order to solve the basic equations, giant electronic computers now produce prognoses of the field of horizontal and vertical motion on a routine daily basis. The results are becoming increasingly more accurate than those produced by the older, more subjective methods. And although meteorologists are still a long way from producing an almanac which will predict the weather for months in advance, prognoses made from the basic physical equations by means of the electronic computer represent a major step ahead in the science.

Weather Forecasts

Most of the meteorological analysis and prognosis described in the previous sections is accomplished by a large central organization. The result, which is generally a combined effort of the professional meteorologist and the electronic computer, takes the form of charts showing the future patterns of horizontal and vertical motion and the positions and intensities of such weather-map features as lows, highs, fronts, and jet streams. From these and other data that represent the large-scale characteristics of the atmosphere, and with a knowledge of the local, small-scale modifications that are likely, the weather forecast is made.

An initial step is the use of highly idealized models of atmospheric phenomena; one of the most basic is the wave cyclone discussed in Chapter 6. In general, a locality in the area up to several hundred miles ahead of the warm front will experience southerly winds (Northern Hemisphere), cool temperatures, and steady precipitation. In the area between the cold and warm fronts, winds will be west or southwest, temperatures will be warm, and there will be little or no precipitation. Behind the cold front, northwesterly winds will prevail, accompanied by cold temperatures and showery type of precipitation. This ideal relationship between the frontal pattern and the weather becomes more complex as the occlusion process takes place. Figure 8-13

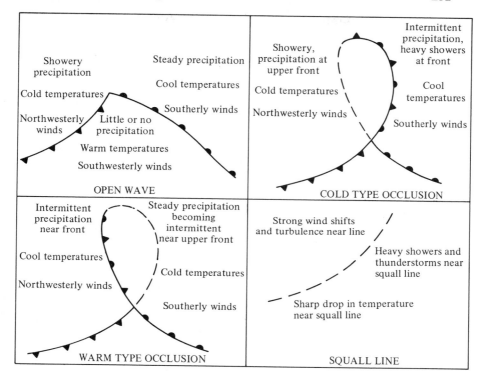

FIG. 8-13. Typical weather patterns associated with fronts.

summarizes some of the more common relationships between fronts and the weather.

Of course, these relationships are greatly simplified and many variations from the ideal patterns occur. In actual practice, the meteorologist must consider the factors that are likely to produce modifications in the ideal models he has used as a first approximation to the weather forecast. For example, when a low-pressure system with its associated fronts is predicted to move over the Nevada Plateau, the precipitation it produces will be generally light because the air masses which reach that region are comparatively dry. Even the air which flows into Nevada from the Pacific Ocean loses much of its moisture as it rises over the Coast Range and Sierra Nevada of California. However, as the same storm system moves eastward into the Mississippi Valley, warm moist air is fed into the circulation from the Gulf of Mexico and the resulting precipitation may be quite heavy. Thus, the forecaster must acquire a basic knowledge of the locality for which the weather

forecast is to be made, and must relate that knowledge to the meteorological prognosis in order to produce the weather forecast.

With a knowledge of the locality and the meteorological prognosis,
many of the basic principles which have been discussed previously in
this text are useful for making a weather forecast. It has been shown,
for example, that instability in the atmosphere is favorable for the
occurrence of thunderstorms; thus, a prognosis which shows cold air
aloft being forced over relatively warm air at the surface will call for
a prediction of showers and thunderstorms. Conversely, warm air
which is subsiding from higher levels over colder air at the surface
will produce a temperature inversion, with its resulting increase in
smog and restricted visibility. To the interested student of the atmosphere, an investigation of the interrelationships among these physical
principles can form the basis for a fascinating study of scientific
weather prediction.

Local weather indications. While accurate weather forecasts require professional training in the science of the atmosphere, as well as
access to meteorological computations of considerable complexity,
some idea of the future weather can be obtained from local indications.
The primary tools required are a good barometer and a means of determining the wind direction free of local obstructions. The latter may
be achieved by observing a windvane or flag on a tall building, or the
motions of lower clouds. Table 8-1 is a summary prepared by the
United States Weather Bureau, showing general relationships between
wind directions and readings of the barometer applicable to much of
the United States. In certain areas, however, these rules should be used
with caution since local influences may be sufficiently strong to out-
weigh the large-scale phenomena upon which the indications are based.

In addition to the relationships given in Table 8-1, other local
weather characteristics may be used to confirm or reject these preliminary indications. For example, the appearance of high cirrus clouds,
followed by cirrostratus and the typical solar or lunar halo, may
presage the approach of a warm front and steady precipitation. Cumulus clouds, appearing during the early morning and beginning to
develop vertically, usually denote increasing instability in the atmosphere, and may be followed by showers and thunderstorms later in
the day. However, if the cumulus clouds are small and do not appear
until afternoon, with no indication of vertical development, fair
weather may generally be expected to prevail. By continually observing the weather, and relating the local indications to the physical
processes which we know to be going on in the atmosphere, many other

TABLE 8-1

Local Weather Indications, Based on Wind and Barometer Observations

Wind direction change	Sea level barometric pressure	Indicated weather
SW to NW	30.10 to 30.20 and steady	Fair, with slight temperature changes for 1 to 2 days.
SW to NW	30.10 to 30.20 and rising rapidly	Fair, followed within 2 days by rain.
SW to NW	30.20 and above and stationary	Continued fair, with no decided temperature change.
SW to NW	30.20 and above and falling slowly	Slowly rising temperature and fair for 2 days.
S to SE	30.10 to 30.20 and falling slowly	Rain within 24 hours.
S to SE	30.10 to 30.20 and falling rapidly	Wind increasing in force, with rain within 12 to 24 hours.
SE to NE	30.10 to 30.20 and falling slowly	Rain in 12 to 18 hours.
SE to NE	30.10 to 30.20 and falling rapidly	Increasing wind, and rain within 12 hours.
E to NE	30.10 and above and falling slowly	In summer, with light winds, rain may not fall for several days. In winter, rain within 24 hours.
E to NE	30.10 and above and falling rapidly	In summer, rain probable within 12 to 24 hours. In winter, rain or snow with increasing winds will often set in when the barometer begins to fall and the wind sets in from the NE.
SE to NE	30.00 or below and falling slowly	Rain will continue 1 to 2 days.
SE to NE	30.00 or below and falling rapidly	Rain, with high wind, followed within 36 hours by clearing, and in winter by colder.
S to SW	30.00 or below and rising slowly	Clearing within a few hours, and fair for several days.
S to E	29.80 or below and falling rapidly	Severe storm imminent, followed within 24 hours by clearing, and in winter by colder.
E to N	29.80 or below and falling rapidly	Severe NE gale and heavy precipitation; in winter, heavy snow followed by a cold wave.
Going to W	29.80 or below and rising rapidly	Clearing and colder.

useful forecasting rules can be developed. One should be careful, however, not to establish general conclusions based on one observation alone; only well-documented observations made over a considerable period can provide relationships which will prove useful in the future.

Local weather observations may also be used to estimate or predict certain characteristics of the atmosphere which may be difficult, or impossible, to observe directly. For example, under certain conditions a simple measurement using a psychrometer to obtain the surface temperature and dew point can provide a good estimate of the height of convective cloud bases. It will be recalled that in Chapter 4 it was suggested that when vertical motion takes place over short periods (a few hours or less), the resulting temperature changes in the rising air generally can be assumed to be adiabatic. Since a good approximation of the dry adiabatic lapse rate is 5.5°F/1000 feet, the temperature, T_h, at any height, h, when the rising air is unsaturated can be determined from the formula

$$T_h = T_0 - (5.5/1000)h \qquad (8\text{-}1)$$

where T_0 is the temperature at the surface, and temperatures are measured in °F and heights are in feet.

In a similar manner, since the lapse rate of dew point in unsaturated air is 1.1°F/1000 feet, the dew point, D_h, at any height, h, is given by

$$D_h = D_0 - (1.1/1000)h \qquad (8\text{-}2)$$

where D_0 is the dew point at the surface.

Now, if a convective cloud is formed by air rising adiabatically from the surface, the base of the cloud will form at the level at which condensation occurs. This, in turn, will be the point at which the temperature and dew point coincide. Consequently, we may let $T_h = D_h$ or,

$$T_0 - (5.5/1000)\, h = D_0 - (1.1/1000)h \qquad (8\text{-}3)$$

With a little algebra, this reduces to

$$h = 227(T_0 - D_0) \qquad (8\text{-}4)$$

This simple formula provides a means for estimating the height of a cloud base. For example, if the surface temperature and dew point are 70°F and 50°F, respectively, a solution of Equation (8-4) gives $h = 4540$ feet as the height of a cumulus cloud which may be forming in air rising from the surface.

Within the cloud itself, the air is saturated, and a rough approximation of the moist adiabatic lapse rate at moderate levels in the troposphere is 3.0°F/1000 feet. Thus, the temperature, T'_h, in the cloud at any height, h', *above the cloud base*, is

$$T'_h = T_h - (3.0°F/1000)h' \tag{8-5}$$

As an example of the use of this formula, suppose it is desired to estimate the height of the "freezing level" in the cloud whose base we have just computed. From Equation (8-1), the temperature of the cloud base is $T_h = 70 - (5.5/1000)4540 = 45°F$. Since we know that the temperature of freezing is 32°F, we substitute this value for T'_h in Equation (8-5). Thus, $32 = 45 - (3.0/1000)h'$. Solving this expression for h' gives 4330 feet, approximately. Adding this value to the height of the cloud base, we have $4540 + 4330 = 8870$ feet for the height of the freezing level above the surface.

Estimates of this kind can be very useful to a pilot taking off from a small airport where no regular weather service is available.

Extended and long-range forecasts. For many operations, a foreknowledge of the weather weeks or months in the future would be very useful. Such activities as planting of crops, building construction, and the like are greatly dependent upon the weather, and many losses could be avoided if major storms, cold waves, and other destructive phenomena could be predicted well in advance. While specific forecasts of this kind are not yet possible, progress is being made. At present, generalized predictions of the weather for periods up to a week (extended-range forecasts), and average deviations of the weather from normal up to one month (long-range forecasts), are about all that can be accomplished.

Extended-range and long-range forecasting commonly makes use of the slowly changing, very large-scale features of the atmospheric circulation. By averaging the daily charts at various levels over several days, the smaller, fast-moving "eddies" and waves in the flow are suppressed and only the persistent large-scale flow remains. At middle and high latitudes, these large undulations of the averaged flow determine to a great extent the general character of the weather over periods up to a few weeks; but the smaller-scale migratory cyclones, which produce the day-to-day weather, remain unpredictable for periods of more than a day or two.

For forecasts beyond one month, all present methods are based upon statistical analyses of the past weather records. One of these includes a search for *periodicities* (cycles) in weather elements. However, aside from the well-known daily and annual cycles caused by the earth's movements, none seems as yet to be very reliable for forecasting the future weather. Another technique involves the use of *analogues*. This consists of looking for patterns in past weather charts that resemble the present one, and then forecasting the present situation to evolve in the future in the same way as did the analogous case in the past. Beyond a day or two, this method is also generally unsuccessful.

Where an operation requires that plans be made for months or years in advance, useful weather information can be obtained from climatological data. Such information as the average and extremes of weather conditions for past years, as well as the frequency of occurrence of operationally critical conditions, can provide helpful advice for all kinds of long-range planning.

8-3. Forecast Accuracy

While some general statements can be made concerning the limitations upon forecasting, it may be useful to examine methods by which the student can verify the validity of claims concerning current accuracy of, or future improvements in, weather forecasts. A useful device for verifying forecasts of any kind is to compare their accuracy with that of forecasts using a simple procedure which requires no skill. Then the forecast being verified can be said to have skill if it achieves a measure of accuracy exceeding that of the "no-skill predictions." A "skill-score" formula for making such an assessment may thus be written:

$$\text{Skill score} = \frac{\begin{pmatrix}\text{number of correct}\\\text{meteorological forecasts}\end{pmatrix} \text{ minus } \begin{pmatrix}\text{number of correct}\\\text{"no-skill" forecasts}\end{pmatrix}}{\begin{pmatrix}\text{total number of}\\\text{forecasts made}\end{pmatrix} \text{ minus } \begin{pmatrix}\text{number of correct}\\\text{"no-skill" forecasts}\end{pmatrix}}$$

Since this formula subtracts the "no-skill" forecasts from both numerator and denominator, the resulting skill score will be positive if the meteorological forecasts are more accurate than those with no skill, and will be zero or negative if meteorological forecasts have no skill at all.

Now, in general, if one verifies weather forecasts made for different periods, using almost any measure of accuracy desired, the result with the present state of the science will be roughly that shown schematically in Figure 8-14. For very short periods (less than a few hours), a forecast of "persistence" of the current weather is generally very diffi-

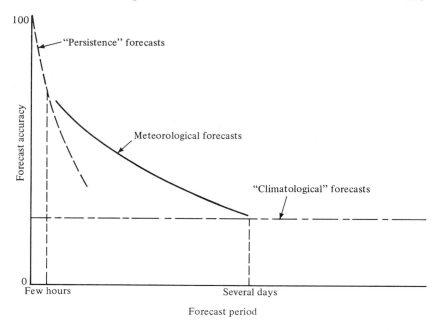

FIG. 8-14. Accuracy of meteorological forecasts compared with "persistence" and "climatology" as a function of forecast period.

cult to improve upon. This is because the network of stations used to observe the atmosphere is usually so coarse that significant weather phenomena may exist between stations and, since such weather may go unobserved, it may also go unpredicted. The use of radar observations, so called "mesoscale-networks" of observing stations, or the like may make such short-period forecasts possible, and some skill over persistence may also be achieved by considering typical weather changes associated with the daily solar cycle; e.g., by forecasting that the temperature will rise during the morning hours and fall during the evening. However, the improvement over persistence is usually very small. Thus for short periods of less than a few hours, a simple prediction that the current weather will persist may be used as an effective "no-skill" forecast.

For periods beyond a few hours, the accuracy of persistence forecasts decays much more rapidly than that of meteorological predictions, so that the latter generally have considerably more skill than the former. As the forecast period lengthens, however, the meteorological forecast accuracy decreases until it approaches "climatology," a term which means in this context that the predicted weather events are distributed randomly (by chance) in the same ratio as they have been observed in the past. Thus, the climatological expectation is a convenient "no-skill" base for the verification of long-period forecasts. Specific meteorological predictions for periods of more than several days in advance are generally not more accurate than "climatology."

Suppose, to choose a typical example, one wishes to examine the validity of rainfall forecasts which have been made by an almanac for a year in advance. For each day of the year, the prediction of rain or no rain is compared with the observed conditions (rain or no rain), and summarized as shown in Table 8-2.

TABLE 8-2

Verification of Daily Rainfall Forecasts Made for a Year in Advance for a Given Location. (Figures in the Table Are Frequency of Occurrence During the Year.)

		Forecast		
		No rain	*Rain*	*Total*
Observed	*No rain*	170	90	260
	Rain	70	35	105
	Total	240	125	365

From this table, it will be seen that the number of correct forecasts is $170 + 35 = 205$. The percentage correct is, therefore, $(205/365) \times 100 = 56$ per cent, which may seem to represent a fairly high accuracy considering that the forecasts were made for an entire year in advance. However, let us compare this result with a "climatological" prediction. Such predictions can be made by distributing, by chance, the forecasts in the same ratio as the observed totals; i.e., if the 240 no-rain forecasts were made by chance, with a knowledge only of the climatology of the observed totals, it would be expected that $260/365$ of these no-rain forecasts would be correct. Thus $240 \times (260/365) = 171$ forecasts of no rain would have been correct due entirely to chance. In the same manner, $125 \times (105/365) = 36$ forecasts of rain would have been correct as a result of chance. Entering these values in the skill-score formula, we have:

$$\text{Skill score} = \frac{(170 + 35) - (171 + 36)}{365 \quad - (171 + 36)} = -.013$$

Since the skill score is slightly negative, it may be concluded that the forecasts summarized in Table 8-2 have no skill above chance.

The chance forecasts may be made by another method which, although somewhat more laborious, may intuitively be more appealing. First of all, a list of the normal number of rainy days for the locality is obtained from the nearest weather bureau office. Then, using a pack of ordinary playing cards, a deck is assembled that has as many black cards as normal rainy days, and as many red cards as no-rain days during the period of interest. The cards are then shuffled and dealt out one by one, the first card being the forecast for the first day of the period, the second card the prediction for the second day, and so on. The result will thus be a "Monte-Carlo" forecast, which may be verified in the same manner as the actual forecasts, and the number of correct predictions used as the "no-skill" forecasts in the skill score formula. If enough days have been used in the experiment, the results of both methods should be approximately the same.

PROBLEMS

Problems marked with an asterisk (*) are the most challenging.

1. What is meant by a "synoptic" weather map? How does the use of the word "synoptic" by meteorologists differ from that given by the standard dictionary definition?

2. Plot the following weather reports at the appropriate locations on a blank map of the United States. Then draw isobars at intervals of 4 mb; i.e., 1004, 1008, etc. Also, locate the fronts by considering the temperatures and wind directions.

Location	Barometric pressure (mb)	Wind direction and speed (kts)	Temperature (°F)
Denver, Colo.	1020	W–15	40
North Platte, Neb.	1015	NW–15	42
Amarillo, Texas	1019	NW–15	45
Oklahoma City, Okla.	1015	N–20	45
Minneapolis, Minn.	1010	N–15	40
Kansas City, Mo.	1011	N–15	44
Shreveport, La.	1011	NW–10	46
Sault Ste. Marie, Mich.	1009	NE–15	40
Chicago, Ill.	1004	NE–20	40
St. Louis, Mo.	1006	NW–20	42
Memphis, Tenn.	1007	NW–15	48
New Orleans, La.	1010	SW–15	76
Cincinnati, Ohio	1002	SW–20	70
Atlanta, Ga.	1008	SW–15	72
Pittsburg, Pa.	1004	SE–20	48
New York, N.Y.	1010	SE–15	46
Norfolk, Va.	1009	SW–15	70
Charleston, S.C.	1012	SW–15	76
Jacksonville, Fla.	1014	SW–10	75

3. Considering your analysis of the weather data of Problem 2, what is the probable current weather at Pittsburg (e.g., clear, raining, snowing, etc.)? What is the probable current weather at Kansas City? At Denver?

4. Again from the analysis of Problem 2, if the storm center is moving northeastward at 20 mph, what will be the predicted weather for Cincinnati tomorrow at this time? What will be the temperature and wind direction?

5. Why do hurricanes which enter the United States along the coast of the Gulf of Mexico usually move northeastward over the eastern seaboard and out into the Atlantic? Why do such hurricanes normally decrease in intensity as soon as they move northward over the land?

6. Why is a "ring around the moon" a fairly good indication of an approaching storm? Can you think of a reason why the following weather rhyme might have a sound meteorological basis?

> "Rainbow in the morning, sailor's warning
> Rainbow at night, sailor's delight."

*7. In addition to Newton's second law of motion, can you name other fundamental laws or scientific principles which might be useful in order to make a prognosis by using an electronic computer? (Hint: What property of the atmosphere, under certain conditions, is an internal source of heat energy? What processes take place without exchange of heat?)

8. If a parcel of air, initially traveling toward the east at a speed of 20 mph, is subjected to an eastward acceleration of 10 mph every hour, what will be its speed at the end of three hours? How far from its starting point will it be at the end of three hours?

9. During the middle of the afternoon on a quiet summer day, the temperature of the air is 82°F, the dew point is 57°F, and cumulus clouds are observed. How high are the bases of the clouds? What is the temperature at the cloud bases? What would be the height of the freezing level in the clouds (assuming that they extend that high)?

10. Assume that the air on a plateau 2000 feet above sea level has a temperature of 50°F, and the dew point is 37°F. If the air flows over a mountain 10,000 feet high and then descends on the other side, at what height will condensation begin? What will be the temperature when the air has descended to sea level on the other side of the mountain?

*11. Obtain the normal number of days each month with precipitation for your locality from the Weather Bureau. Make a "Monte-Carlo" forecast for each of the next several months by using a deck of playing cards, in the way described on pp. 258-59. Using this as the "no-skill" prediction, compute the skill score for (a) an almanac forecast, and (b) the forecasts published in the daily newspaper. Which forecasts show the greater skill?

12. By observing the weather (winds, clouds, etc.) at 8:00 A.M. each morning, make a forecast concerning whether or not you expect it to be raining at noon that same day. At the end of a month or two, compare your forecasts with "persistence" (i.e., predicting that the weather at noon will be the same as that at 8:00 A.M.) using the skill-score formula. Were your predictions more skillful than persistence?

13. Identify the major crests and valleys at 500 mb in Figure 5-3. How do the sea level cyclones and anticyclones move in relation to the large-scale wave patterns?

14. Is the complexity of the frontal analyses on the sea level charts greater over the oceans or over the continents? Why should there be a difference?

15. Examine the latest weather map published in your local newspaper. Predict whether there will be precipitation (and, if so, the type) and what the temperature, wind, and cloudiness will be in your city 24 hours from the time of the map. List the factors you took into account in predicting each element.

9

Applications of Meteorology

9-1. Economic Impact of the Weather

As the technological complexity of modern society
grows, the influence of such environmental factors as
the weather becomes increasingly important. For example, a generation ago a severe storm which resulted
in the loss of electric power in the home for a day or
two would have been considered no more than an annoying inconvenience; today, in many homes, there
would be no light, the furnace would not operate, it
would be impossible to cook a meal, food in the re-
frigerator and freezer would spoil, and there would be
no hot water.

While the ordinary tasks of daily life may thus be affected significantly by the occurrence of a single adverse weather event, modern agricultural, business, and industrial activities may be influenced by even relatively small changes in the weather. Planting, cultivating, harvesting, and marketing of crops are increasingly dependent upon information concerning the past and future weather. Efficient operation and control of both surface and air transportation requires accurate information concerning storms and other atmospheric hazards. Growing industrialization increases the need for more precise weather data for planning factory locations and for day-to-day plant operation. Even such fundamental problems as efficient utilization of our basic resources (e.g., water, petroleum deposits, natural gas, as well as the atmosphere itself) are dependent upon a knowledge of the meteorological factors which affect their use.

To some extent, the technological progress associated with our growing society has made it possible for the meteorologist to solve some of his own scientific problems. Thus, somewhat paradoxically, the electronic computer and meteorological satellites are products of an industrial age which is creating the complex problems which the computer and satellite are helping to solve.

It should also be noted, however, that one important goal of modern technology is to design operations which will be independent of the weather. Many examples of progress toward this goal may be cited. For instance, although early railroads frequently had to shut down during severe storms, the modern railway is now rarely affected by the weather. Similar improvements are being made in aviation, and all-weather flying is clearly the ultimate aim of the aviation industry. In agriculture, also, such developments as hybrid corn and drought-resistant wheat have helped to compensate for certain adverse effects of the environment. An important contribution of the meteorologist is in providing the basic knowledge about the atmosphere required for the continuance of such progress.

In this chapter we shall examine some of the economic and other consequences of the more practical applications of meteorology.

9-2. Agriculture

From the beginning of man's efforts to till the soil, weather has been recognized as exerting a paramount influence on crop production. As early as 2000 b.c., the Babylonians, concerned with the effects of weather on the agriculture of the Euphrates and Tigris Valleys, made studies based on a supposed relationship between astronomical phenomena and the weather. One of the earliest weather forecasts, signed with the forecaster's name, was addressed to the king: "When it thunders on the day of the moon's disappearance, the crops will prosper and the market will be steady. /s/Asaridae." More than a thousand years later, the Egyptians made serious attempts to learn the causes for the annual flooding of the Nile River. Not until the explorations of Dr. David Livingstone in the nineteenth century, however, was it determined that monsoon rains falling in the headwaters of the Nile were responsible for the almost unfailing regularity of the high water which brought rich topsoil to the river delta where the crops were grown.

In recent years, the explosion of the world's population has re-emphasized the need for increased crop production in order to alleviate problems of hunger and starvation. Such improvements as better seed, modern farm machinery, and more effective pest control must be employed for this task. However, all of these devices are affected by the atmospheric environment, and their use can be made more efficient through an adequate consideration of weather and climate.

Plant growth. The relationship between plant and animal growth and the weather is extremely complex, involving a great many factors. This is to be expected, when one considers that living matter depends for its survival on a continual exchange of energy (through radiation, conduction, and convection) and of mass (water, oxygen, carbon dioxide, etc.) with its environment. However, to simplify the discussion, we shall confine ourselves to just two weather elements that are easy to measure and are known to have very significant effects, especially on plant life: temperature and precipitation.

Figure 9-1 is a simplified diagram showing the general dependence of natural vegetation on the annual values of temperature and precipitation. In general, forests grow naturally where the rainfall is high, and trees can also exist with relatively low temperatures. Deserts are found, regardless of temperature, if the rainfall is below about 10 inches per year. In between are the grasslands which, in cultivated areas, are productive of much of the cereal foods used by man. Of course other important factors, such as the seasonal variation of temperature and rainfall, should also be considered.

Through the processes of natural evolution, the broad spectrum of plant life has become adapted to a wide range of environmental conditions. At the extremes, certain algae grow and flourish in hot springs where the water is near the boiling point, while arctic plants can survive temperatures as low as −90°F. In the desert regions, a class of

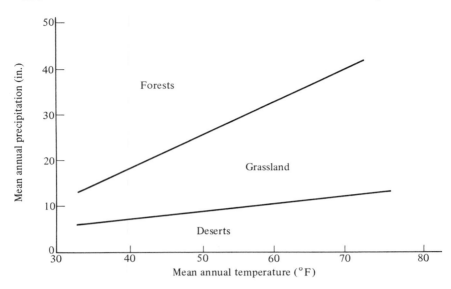

FIG. 9-1. Dependence of natural vegetation on temperature and precipitation. (After Smith.)

plants called Xerophytes can survive with little water, and certain flowers produce seeds which may remain dormant for years until rainfall occurs and then burst forth in a dramatic display of color on the desert floor. However, most plants will grow only within a relatively narrow range of temperatures, from about 32°F to about 122°F, and they require a minimum of about 10 inches of precipitation.

Temperature and Crops. For each species and variety of plant, there exists a certain combination of maximum and minimum temperatures beyond which the plant will not grow, while between these limits is a temperature at which the growth is optimum. These three temperatures are called "cardinal growth temperatures." Table 9-1 shows such temperatures for a few typical crops.

The cardinal temperatures for growth vary, not only among plant varieties, but also with the stage of development of a particular plant. For example, a plant in its initial stages of growth usually has a lower optimum growth temperature than it has when near maturity. Furthermore, an optimum growth may not be the most favorable from the standpoint of crop production; it may result in plants that are susceptible to disease, frost damage, or damage by wind.

TABLE 9-1

Cardinal Growth Temperatures for Typical Crops

Crop	Minimum (°F)	Optimum (°F)	Maximum (°F)
Cool season; e.g., oats, wheat, barley	32–41	77–88	88–99
Hot season; e.g., melons, sorghums	59–64	88–99	111–122
Truck crops; e.g., beets, carrots, lettuce	34–40	50–85	85–100

In an attempt to devise a practical method for determining the date at which crops will be ready for harvest, a "heat-unit" system is used by many agriculturists. In this procedure, the amount by which the daily mean temperature exceeds some base temperature (usually the approximate minimum cardinal growth temperature) is cumulated during the growing season. For example, if the base temperature for a particular plant is 35°F, and the mean temperature on a given day is 50°F, the number of heat units (usually called "degree days") is 15. Beginning with the start of the growth season, these daily values are added; then, if 1000 degree days are required for that crop to mature, it should be ready for harvest by the time 1000 degree days have been accumulated. Table 9-2 summarizes the base temperatures and number of degree days to maturity for a number of crops.

TABLE 9-2

Estimated Heat Units for Certain Agricultural Crops to Reach Maturity

Crop (variety, location)	Base temperature (°F)	Heat units to maturity (degree days)
Beans (Snap, S. Carolina)	50	1200–1300
Corn (Sweet, Indiana)	50	2200–2800
Corn (Golden Bantam, S. Carolina)	50	1400–1500
Cotton (Delta Smooth Leaf, Arkansas)	60	1900–2500
Peas (Early, Indiana)	40	1100–1200
Peas (Medium or Late, Indiana)	40	1400–1600
Peas (Alsweet, Wisconsin)	40	1300–1400
Peas (Perfection, Wisconsin)	40	1700–1800
Rice (Vegold, Arkansas)	60	1700–2100
Rice (Bluebonnet, Arkansas)	60	2400–2600
Wheat (Indiana)	40	2100–2400

The heat-unit system is in wide use by farmers who grow crops for commercial canning. By using the average degree-day figure for each day during the growing season at a given place, the dates can be determined on which certain crops should be planted in order to provide for an orderly flow of mature crops to the cannery. Other uses include the scheduling of pest control operations and the selection of suitable crops to be planted in a given area. It will be noted that many factors pertinent to plant growth are not included in the heat-unit system; e.g., the amount and timing of precipitation and the amount of sunlight. However, the system provides a simple approximation for practical use.

The protection of crops from damage due to extreme temperatures represents a field in which the meteorologist has been called upon to provide a great deal of assistance. In the use of insecticides and fungicides for pest and disease control, for example, high temperatures within a day or two following treatment may cause severe damage. Thus, the decision to spray an orchard on a given day may depend on the temperature forecast. But it is in the protection of crops from frost and freezing temperatures that the farmer places some of his greatest dependence on meteorological information.

Freezing temperatures may occur either with, or without, the occurrence of frost. In either case, freezing and consequent rupture of plant cells may occur. The extent of the damage suffered by the plant depends upon how long low temperatures persist, the general health of the plant, and the weather preceding and following the freezing. If warm, sunny weather prevails prior to the frost, so that the plant is growing rapidly, damage is likely to be greater than if the preceding weather has been cool and cloudy.

The cold that can be endured without damage is dependent upon the plant species. Table 9-3 gives, for a number of common plants, the temperature which can be endured for 30 minutes or less without material damage.

Crops can be protected against frost or freezing temperature in a number of ways. Since the cooling of the earth's surface at night is often due primarily to radiational heat loss, one method of frost protection is to reduce the heat loss. The familiar greenhouse serves this purpose. The glass covering permits the incoming short-wave solar radiation to pass through during the day and heat the ground and plants, but it inhibits the outward passage of long-wave radiation from the interior of the greenhouse, and mixing of the warm air inside with the colder outside air, thus conserving the interior heat. In principle, this method may be used for the protection of field crops. This is accomplished by such devices as "hot caps"—translucent plastic or waxed

TABLE 9-3

Temperatures (°F) Endured for 30 Minutes or Less by Certain Fruits and Vegetables. These Are Average Values, and Some Variation May Be Expected, Depending on the Health of the Plant and Other Factors.
(After Young)

Variety	Deciduous fruits		
	Buds closed but showing color	Full bloom	Small green fruit
Apples	25	28	29
Apricots	25	28	31
Cherries	28	28	30
Peaches	25	27	30
Pears	25	28	30
Plums	25	28	30
Prunes	23	27	30
Walnuts (English)	30	30	30

Variety	Citrus fruits		
	Buds or blossoms	Small green fruit	Mature fruit
Grapefruit	30	29	28
Lemons	30	30	29
Oranges	30	29	28

Citrus trees will withstand a temperature of 22–25° without serious damage.

Variety	Vegetables
	Nature of damage by frost
Carrots	Tops "burned" when temperature reaches 23°F, but roots not affected.
Cauliflower	Discolored after temperature reaches 25°F.
Cucumbers	High percentage killed by 31°F for half hour.
Lettuce	Outside wrapper leaves damaged by 24°F.
Peas	Bloom killed by 30°F; small pods damaged by 29°F.
Squash	Small vegetables damaged by 26°F.
Sweet potatoes	High percentage killed by 31°F for half hour.

paper cups which are inverted over individual small plants, or by long sheets of plastic which are placed over semicircular wire supports covering each row of plants. Such devices not only provide some frost protection at night, but the higher temperature they engender during the day promotes more rapid maturing of the crops. Other procedures which conserve heat include the use of lath or cloth screens, and piling brush or soil over the plants.

Another basic method of frost protection is to replace the radiational loss by the addition of heat. The most effective procedure involves the use of orchard heaters which burn liquid or solid petroleum fuel. Early forms of such heaters were simply open pails filled with oil. Because of the dense smoke which accompanied the burning oil, they were colloquially called "smudge pots," and it was believed by many farmers that the smoke produced a blanket which inhibited the loss of heat. However, tests have shown that this effect is negligible, and that the primary protection is provided by the heat released by combustion of the oil. Since such heat is most efficient when a minimum of smoke is produced, and also because of the necessity to avoid polluting the air, most heaters (Figure 9-2) are now designed to emit as little smoke as possible.

The effectiveness of heating is due to the presence of a temperature inversion which normally accompanies nocturnal radiational cooling. Hot gases from the heaters are mixed by turbulence with the surrounding cold surface air, so that the temperature of the mixture is only a few degrees above its initial state. The warmed air rises (cooling adiabatically) until it reaches a level within the inversion with the same temperature; the upward movement then gradually slows down and finally stops. This is illustrated in Figure 9-3; note that the heating effect only extends upward slightly over 50 feet, and that the principal heating is from the surface to the tops of the trees in the orchard. Accordingly, on a normal frosty night it is not necessary to heat "all of outdoors"; because of the inversion, only the lowest layers of the atmosphere need to be heated.

The presence of the warm air in the temperature inversion suggests that, if this source of heat could be brought down to the ground and mixed with the surface air, the resulting temperature in the orchard would be raised. This can sometimes be accomplished by using "wind machines"; i.e., power-driven fans placed on high towers within the orchard (Figure 9-4). The turbulence generated by the wind machine causes the warm air aloft to mix with the cold surface air and, on nights when an adequate temperature inversion exists, the temperature within the growing area can be raised several degrees. Figure 9-5 shows the temperature change observed in an orchard with the use of such a machine. On this occasion, the temperature at an elevation of 50 feet was only 6°F higher than at the surface, but the surface air was warmed more than 3°F after only 35 minutes operation of the wind machine.

Occasionally, below freezing temperature is caused primarily by an invasion of cold air rather than by nocturnal radiation. Such situations are called "freezes" to distinguish them from the radiational

FIG. 9-2. Orchard heater. (Courtesy of Hauter & Associates.)

"frosts." Under freeze conditions, there is, of course, no warm air aloft (no temperature inversion) so that the task of heating an orchard or other farm crop becomes much more difficult; also, the operation of

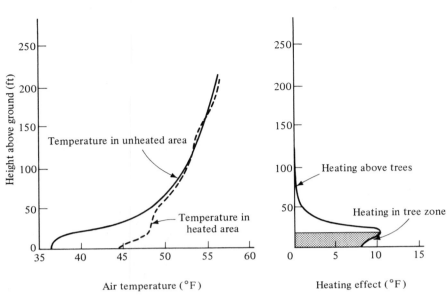

FIG. 9-3. The effect of orchard heating. (Thirty-three gal of fuel per hour burned in a 15-acre citrus orchard, 45 heaters per acre. After Kepner.)

wind machines under freeze conditions merely mixes the cold air at the surface with colder air aloft. Fortunately, such conditions occur rarely in agricultural areas during the growing season.

Because of its importance, specialized weather services have been developed to assist growers and marketers in reducing frost damage. This service includes forecasts of anticipated periods of below freezing, expected minimum temperature, advice concerning the strength and height of the temperature inversion, the anticipated rate of fall of temperature, and the probable heating requirements. Following a severe frost or freeze, information concerning the extent of damaging temperature is provided to marketing agencies for use in estimating crop yields.

Water and Crops. Along with temperature, the amount of water needed by plants is the most critical factor which determines their growth. In fact, most plants contain considerably more water than anything else. Mineral elements are transported from the soil in a water solution; also carbon dioxide from the air is dissolved in water and, through the process of photosynthesis, forms the material which provides for plant growth. Thus, crops require a great deal of water—

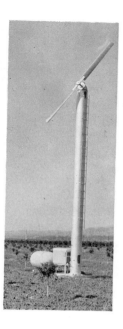

FIG. 9-4.   Orchard wind machine. (Courtesy of Hauter & Associates.)

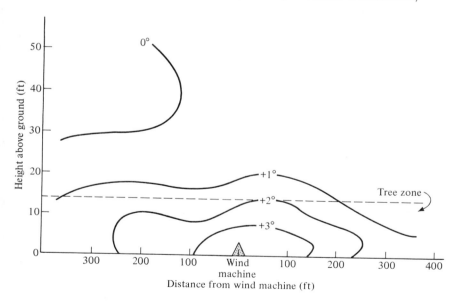

FIG. 9-5.   Temperature change (°F) 35 minutes after start of wind
machine. (Surface temperature inversion of 6°F extending
to 50 ft. After Crawford and Leonard.)

273

about 200 to 300 pounds of water per pound of dry plant material are required for adequate growth of those plants which use water least efficiently, while up to 100 pounds of water per pound of dry material are needed for more efficient crops (see Table 9-4).

The approximate precipitation equivalent of these figures is indicated by the transpiration (evaporation from leaves and other plant surfaces) which takes place as crops grow. In the midwestern United States, transpiration from corn is about the water equivalent of 15 inches of rainfall during a growing season, while from deciduous trees, transpiration may amount to 20 inches or more per growing season. Additional amounts of water are accounted for by evaporation from the soil to the atmosphere, and runoff or percolation into the soil itself. This water must be supplied by precipitation or, where this natural source of moisture is insufficient, by irrigation. Figure 9-6 shows the effect of providing adequate moisture through irrigation of corn plants in a region of normally little rainfall during the growing season.

The amount of irrigation water required for plant growth depends largely upon the nature of the plant, temperature, the amount of sun-

TABLE 9-4

*Comparison of the Efficiency of Various Crops in the Use of Water**
(After Young)

Crop	Water use		Yield	Food value	Water use efficiency	
	in.	1000 gal/acre	1000 lb/acre	100 cal/lb	lb/1000 gal	cal/gal
Wheat	20	543	6.0	14.8	11.1	16.4
Sorghum	27.6	749	8.0	15.1	10.7	16.2
Peanuts	34.5	937	4.0	18.7	4.3	8.0
Dry beans	20.6	559	3.0	15.4	5.4	8.3
Safflower	33.4	907	4.0	14.2	4.4	6.2
Soy beans	33.4	907	3.6	18.3	4.0	7.3
Potatoes	16	434	48	2.79	111.	31.
Tomatoes	19	516	60	0.95	116.	11.
Oranges	53.1	1442	44	1.31	30.5	4.0
Cotton	34.5	937	1.75 (lint) 2.8 (seed)	––	4.9	––

* Values apply to a particular region of the world and they may vary considerably from place to place.

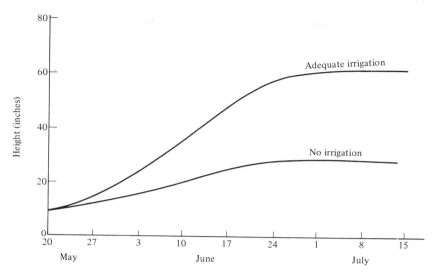

FIG. 9-6. Effect of irrigation on corn. (After Kramer.)

shine, and the natural precipitation which occurs during the growing season. An approximate relationship which includes these factors can be expressed quantitatively by the formula (adapted from Wang):

$$W = k \; (P_1T_1 + P_2T_2 + P_3T_3 + \ldots + P_nT_n) - R \qquad (9\text{-}1)$$

where

$W =$ the amount of irrigation water, in inches, required during a growing season of n months;

$k =$ a coefficient, the value of which depends upon the crop (Table 9-5);

$P_1, P_2, \ldots P_n =$ ratio of total possible monthly hours of sunshine to total annual hours for each of n months (Table 9-6);

$T_1, T_2, \ldots T_n =$ mean monthly temperature for the region (°F) for each of n months;

$R =$ total growing season rainfall (inches)

As an example, consider the problem of determining the approximate amount of irrigation water required for a vegetable crop to be grown in San Diego County, California (Latitude 32°). The crop is to be planted in early March and harvested in late June. From Appendix 9, the normal temperatures and rainfall during this period are:

	Temperature (°F)	Rainfall (in.)
March	59	1.6
April	58	0.8
May	61	0.2
June	66	0.0
	Total	2.6

Using the appropriate interpolated values of k and P from Tables 9-5 and 9-6, the computations indicated by Equation (9-1) may be carried out as follows:

$$W = 0.6 \,(.08 \times 59 + .09 \times 58 + .09 \times 61 + .10 \times 66) - 2.6 = 10.6 \text{ in.}$$

TABLE 9-5

Value of k for Various Crops (After Blaney and Criddle)

Crop	k	Crop	k
Alfalfa	0.9	Grass hay	0.8
Corn	0.8	Rice	1.2
Cotton	0.7	Vegetables	0.6
Pasture	0.8	Citrus trees	0.6
Potatoes	0.8	Deciduous trees	0.6

TABLE 9-6

Value of P; i.e., Ratio of Total Possible Monthly Hours of Sunshine to Total Possible Annual Hours for Various Latitudes

Latitude:	30°	40°	50°
Jan.–Mar.	.07	.07	.07
Apr.–June	.09	.10	.10
July–Sept.	.09	.10	.10
Oct.–Dec.	.08	.07	.06

Thus, to supplement the amount of moisture normally received, about 10 or 11 inches of irrigation water must be added. A further complication arises because all of this water is not available for use by the plant itself—as mentioned earlier, some moisture is lost by evaporation from the soil, and by runoff and percolation into the soil. The amount which is lost by this means is dependent upon a number of factors, including soil condition, cultivation practices, and the like. In fact, as much as 30 per cent may be lost through these processes, so that a total of 13 or 14 inches of irrigation water may actually be needed in the example given above.

Occasionally, too much water may also be a source of difficulty. Excess moisture may be caused by heavy rainfall, improper irrigation, or even flooding from nearby rivers. The result is frequently lower soil temperature, lack of aeration of root systems, and generally lower crop yield. These problems may be alleviated by proper cultivation, and by grading or building drainage ditches so that the excess moisture will be removed. When flooding occurs, it is usually widespread, so that protective measures normally involve building of large-scale flood control structures.

Weather and farm animals. The weather usually affects animals indirectly through its influence on the growth of plants which are used as feed, but there are direct influences as well. For example, milk production significantly decreases when the air temperature rises above 90°F. High temperatures also result in decreased egg production and a lower rate of reproduction of most farm animals. An increase in the amount of sunlight may improve egg production, but too much sun will cause cattle to cease grazing and seek the shade. On the whole, however, animals appear to be more adaptable to their climatic environment than plants and, because of their mobility, are able to seek shelter when unfavorable weather occurs.

Agriculture today is a rapidly growing science. A great deal of research is continually being carried on by scientists in an effort to more adequately understand the nature of plant and animal life, and its relationship to the atmospheric environment. In the foregoing section, we have only been able to examine a few of the more typical problems. Some other aspects of the subject are described in the references and in basic textbooks on plant and animal physiology.

Forestry

The forests of the world represent natural resources which provide not only timber and other products, but also contribute to the conservation of water, shelter and food for wildlife, and the recreational needs of the public. Many aspects of the environment are involved in the growth of this natural resource, but the weather is most closely related to the forest's greatest single hazard—fire. In the United States alone, forest fire losses are estimated to exceed half a billion dollars annually, and fire detection and suppression costs exceed 100 million dollars each year.

Forest fires may be started by man himself through carelessness, or by such natural causes as lightning. In either case, the spread of a small blaze may take place with astonishing rapidity when trees and underbrush have been dried out by a long period of high temperatures and low humidity. In order to provide a measure of the potential for fire inception and spread, a "fire danger rating" has been developed by combining the related sciences of forestry and meteorology.

The specific nature of the fire danger rating depends somewhat on the region in which it is used, but the primary weather factors include measurements of temperature, moisture (humidity), and wind speed. In most cases, there is also included a factor which measures the moisture content of the potential fuels; i.e. the grass, brush, and trees. This is accomplished in some regions by a subjective assessment of the dryness of the ground cover; for example, if it is green, partially dried out, or completely dead. In other areas, an objective measurement is obtained by use of a "fuel moisture stick" which decreases in weight in direct proportion to the drying out of the forest itself. By combining the weather factors with the measurement of the fuel moisture, the fire danger may be given a numerical rating which, in turn, can be interpreted in such terms as "safe," "moderate danger," "high fire danger," or the like. When the rating is very high, certain areas may be closed to the public, logging and other operations may be stopped, and fire fighting crews may be placed on alert for possible action.

As may be inferred from an examination of the factors involved in the fire danger rating, the spread of a forest fire takes place most rapidly under conditions of high temperature, low humidity, and strong winds. Generally, such situations are associated with foehn winds. The adiabatic heating and resulting decrease in humidity associated with foehn winds produce especially dangerous fire weather conditions.

Frequently, the surface heating caused by a forest fire will bring about strong upward vertical air currents. If the vertical motion extends high enough into the atmosphere, and if there is sufficient moisture in the air, a cumulus cloud will form above the fire. Such a "cap" cloud is shown in 9-7. If the cumulus continues to grow, a thunderstorm may develop which will bring enough rain to put out the fire which caused it! This is not a frequent occurrence, however, and most forest fires must be first controlled and then extinguished by the efforts of fire fighters. Assisting in these efforts are such modern devices as aircraft which release a fire-retardant solution over the fire, and bulldozers which tear out trees and underbrush to build a fire-break. But much of the final defeat of a forest fire must be accomplished by the fire fighter himself using only such weapons as a pick and shovel.

FIG. 9-7. Cumulus cap clouds formed over a forest fire. (Courtesy of U.S. Forest Service.)

9-3. Aviation

Of all weather-service users, the aviator is probably more continuously aware of atmospheric conditions than any other. While persons on the ground are able to seek shelter from many of the adverse effects of the weather—rain, snow, strong winds, and the like—the airplane pilot must either avoid them or make his way through them. Of course, many times the weather is helpful; clear skies, assisting tail winds, and smooth air make flying an enjoyable experience. But the pilot must also be trained to cope with dangerous or critical conditions.

While those who fly need not become expert meteorologists, all licensed pilots must pass certain proficiency tests which include a basic knowledge of the weather. Those who take advantage of their meteorological training not only gain a better understanding of the weather phenomena they encounter in flight, but are also able to operate their aircraft more safely and efficiently. In this section, we shall examine some of the ways in which meteorology may be applied in achieving these goals.

Flying weather—takeoff and landing. While the aircraft is subjected to a number of weather hazards at all times during flight, it is convenient to discuss those atmospheric phenomena which affect the aircraft primarily during takeoff and landing operations separately from the weather which is of most concern while flying enroute. In general, the meteorological elements which are of primary significance during takeoff and landing are (1) restrictions to horizontal and vertical visibility; e.g., fog, clouds, smoke, haze, precipitation; (2) turbulence or sudden wind shifts in the vicinity of the airport or on the runway; and (3) icing on the aircraft or in the engine.

Restrictions to horizontal and vertical visibility are factors which determine whether a pilot may take off and fly under Visual Flight Rules (VFR), or whether he must conduct his flight under Instrument Flight Rules (IFR). Some idea of the relative frequency with which the combination of restricted visibility and ceiling occur in the United States is shown in Figure 9-8. During the summer, restrictions to visibility and ceiling occur principally near the maritime borders of the United States, where the location of large bodies of water is conducive to the formation of fog and low cloudiness. In the winter, middle-latitude storms are the most frequent producers of low visibility and ceiling. However, even here, the effect of moisture from the Atlantic and Pacific Oceans, and the Gulf of Mexico, is evident.

Fog is the most frequent cause of operationally critical reductions in horizontal visibility on the airport. The meteorological processes which bring about fog formation have already been covered in Chapter 3; of these, perhaps radiation fog is the most critical during takeoff and landing, since it tends to be thickest near the runway surface. However, all types of fog are important for aircraft operations. When surface fog occurs, visibility at flight level above the fog is usually good.

Industrial smoke and other pollutants occasionally bring about significant reductions in visibility. When such pollutants occur in combination with fog or low clouds, the resulting "smog" frequently becomes a serious aviation hazard. Other restrictions to visibility include precipitation in the form of drizzle, rain or snow, blowing dust or sand, and drifting snow. Such phenomena occur less frequently in most areas than fog or its combination with smoke, but in some instances they can be a serious hazard.

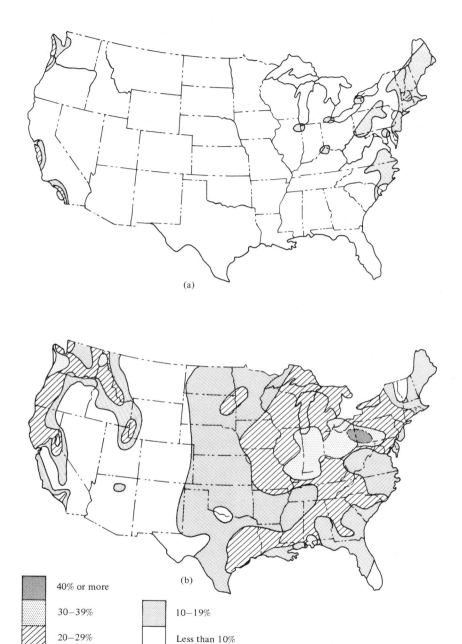

40% or more

30–39%

20–29%

10–19%

Less than 10%

FIG. 9-8. Percentage of hours when the ceiling is below 1000 ft and visibility is less than 3 mi. (a) Summer; (b) Winter.

281

Of the cloud types, stratus cloudiness represents the most critical restriction for takeoff and landing operations. Since fog and stratus clouds frequently occur together, these two weather phenomena usually are responsible for the greatest number of closed airports, and the longest weather-caused traffic delays.

Turbulence is associated with irregular cells or eddies of air, the motions of which cause annoying or even dangerous jolts to an aircraft in flight. While turbulent eddies occur in a wide range of sizes, those which cause the greatest bumpiness are roughly the same size as the aircraft. The frequency of the jolting is also greater as the speed of the aircraft increases, since more turbulent cells are encountered in the same length of time. This may make the turbulence seem more severe. Thus, aircraft of different sizes, flying at different speeds, may experience different degrees of jolting while flying through the same turbulent air.

During takeoff and landing, the primary causes of turbulence are convective currents produced by uneven heating of the earth's surface, and eddies in the wind induced mechanically by the presence of buildings or rough terrain near the airport. The occurrence of larger-size convective currents during landing may result in overshooting or undershooting the airport runway, as illustrated in Figure 9-9. If the total component of the convection currents is upward along the glide path, the plane may overshoot the runway, while if the total component is downward, the plane may land short of the runway. Convective currents can almost always be inferred if cumulus cloudiness is present in the area around the airport; but even if such visual evidence is not apparent, convective currents may be occurring if the ground surface along the path of the aircraft is of uneven composition; i.e., alternating green fields, concrete roadways, rivers or lakes, and barren ground. On a clear warm day, these variations in the ground surface will be associated with irregular absorption of solar radiation, and the consequent differences in heating of the air may cause vigorous, but invisible convection cells or "thermals."

Turbulence caused by the blowing of the wind over and around buildings, hangars, and rough ground surfaces produces a series of irregular and complex eddies which may seriously affect the takeoff and landing of aircraft. These mechanically induced eddies are usually carried along with the wind. The degree of turbulence they produce is generally more intense the greater the wind speed, and more prevalent the greater the atmospheric instability (Figure 9-10). Sharp-edged structures are more apt to cause turbulent eddies, both on the windward and leeward side, than are smooth surfaces such as rolling hills.

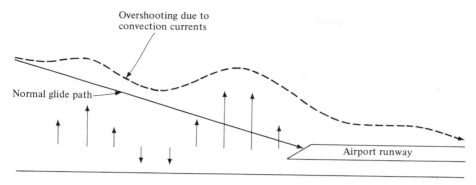

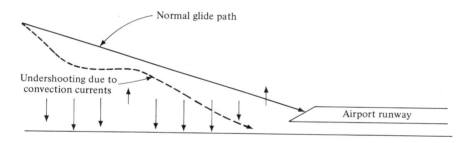

FIG. 9-9. The effect of convective currents on an aircraft landing.

When winds are light, the turbulent eddies tend to remain close to the structures with which they are associated; but when winds are strong, the turbulence may be carried by the wind well out into the runway area.

A phenomenon called "wake turbulence" is associated with mechanical eddies which frequently form in the wake of aircraft in flight. Such turbulence is produced by the airfoils of the moving plane and is much like the eddies around buildings except that it may be considerably more severe. (See Figure 9-11.) Wake turbulence over the airport is most severe if the air near the ground is very stable, and the turbulence persists longest near the runway where it is produced if the air is calm. Under such conditions, the turbulence from a large

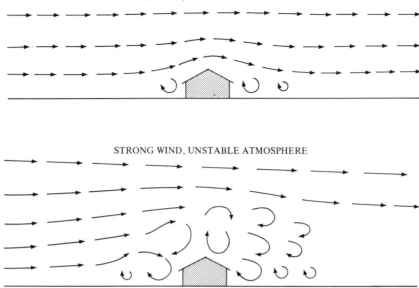

STRONG WIND, UNSTABLE ATMOSPHERE

FIG. 9-10. Turbulence produced by obstructions to the wind.

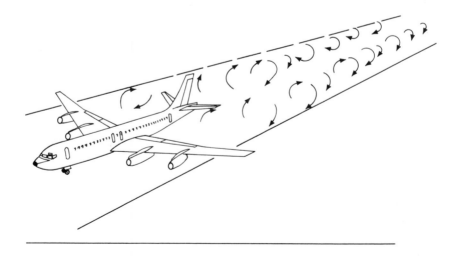

FIG. 9-11. Turbulence in the wake of a landing aircraft.

airplane may be felt as long as five minutes after the aircraft which caused it has passed. Wake turbulence is operationally most critical when a large jet aircraft is followed too closely in takeoff or landing by a light airplane. The turbulence produced by the large plane may be sufficient to cause complete loss of control by the pilot of the following light plane.

Closely related to the problem of turbulence is the occurrence of cross winds or sudden wind shifts during a takeoff or landing. Since an aircraft must take off or land in a straight line along the runway, any wind blowing at an angle to the runway may cause the airplane to overturn or run off the landing surface. Sudden and unexpected shifts in the wind may occur during thunderstorms, with a frontal passage, or in connection with small-scale phenomena such as the leading edge of a sea breeze. Normally, such weather conditions are easily identifiable, and warnings of operationally significant changes in wind are provided to pilots and traffic controllers at large air terminals. However, pilots should always take precautions to observe local meteorological conditions which may bring about a sudden shift in the wind during takeoff or landing.

Aircraft icing is one of aviation's most serious hazards. When it forms on the wings or fuselage it may seriously reduce the flying capability of the airplane; if it forms in the air intake of the engine, it can cause a hazardous reduction in power. Icing may also affect the efficiency of the control mechanisms, the brakes, landing gear, or flight instruments.

Icing on the aircraft structure is generally caused by the freezing of supercooled water droplets when they are struck by the plane in flight. It is most severe when the air temperature is between $0°C$ and $-10°C$, but may be encountered with temperatures as low as $-40°C$. Any type of fog or cloudiness which contains liquid water droplets at these temperatures may produce aircraft icing. However, the most dangerous ice accumulation is usually found in freezing rain, where large amounts of ice may build up in a few minutes, or in cumulus or thunderstorm clouds. Occasionally, also, ice may form in clear air if the atmosphere is close to saturation. In this case, the decrease in pressure over the airfoil may result in sufficient cooling so that sublimation will take place directly on the aircraft.

Structural icing may form as clear ice (glaze), rime ice, or frost. Clear ice is virtually transparent, with a glassy surface similar to the glaze which forms on foliage, fences, and other objects near the ground during freezing rain. It generally occurs in clouds or precipitation with high moisture content, large droplet size, and temperatures

only slightly below freezing. It is therefore most common in cumulus-
type clouds and in freezing rain or drizzle. Although clear ice is usually
smooth and tends to conform to the shape of the airfoil (Figure 9-12),
it is very heavy and difficult to remove even with efficient deicing
equipment.

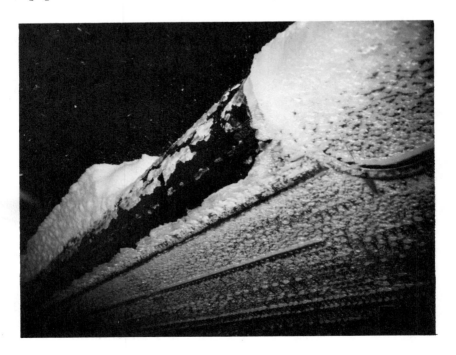

FIG. 9-12. Clear wing icing (right leading edge and underside). (Courtesy of NACA.)

Rime ice is generally opaque; it consists of small granules of ice
which produce a rough irregular surface (Figure 9-13). It generally
occurs when the temperature is between $-10°C$ and $-20°C$. It is most
frequent in stratiform clouds which have a small droplet size, although
occasionally it may also be encountered in cumulus clouds with limited
vertical development. Rime ice has a much greater disturbing effect
on the shape of the airfoil than clear ice, but is relatively easy to remove with modern deicing equipment. It should be noted that the
meteorological conditions which produce the two types of icing rarely
occur separately; consequently, a mixture of clear ice and rime ice will
be encountered in most icing situations.

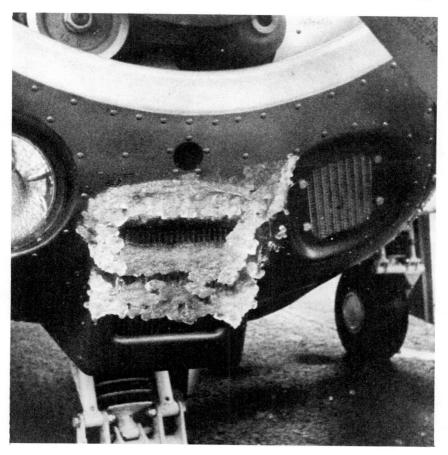

FIG. 9-13.   Rime icing. (Courtesy of Norman Hoffman/*A.O.P.A. Pilot.*)

Frost may form on an airplane if it has been flying in a region of below freezing temperatures and suddenly moves into an air mass with above-freezing temperatures and high relative humidity. Typically, this can occur during a landing operation, when the airplane descends from the cold air of high altitudes to the warmer air near the surface. Under such conditions, the windshield may suddenly be covered by a thin sheet of frost which can restrict the pilot's visibility at a most critical time. Frost may also occur on aircraft which are parked outside at an airfield during a cold night.

Icing may occur in the engine of either conventional or jet aircraft. In the conventional powerplant, the most serious hazard is the formation of ice in the carburetor. Here, the cause of ice formation is cooling due to vaporization of the engine fuel, and expansion of air as it is brought in through the intake manifold. If the relative humidity of the air is high, ice may form in the carburetor even with air temperatures outside the aircraft as high as 25°C. In the case of jet aircraft, cold outside air temperatures may actually cause ice to form in the liquid fuel itself. This is because water is readily absorbed and mixed with the jet fuel when the relative humidity is high.

Protective measures against icing on wing and tail surfaces includes the use of rubber covers (boots) attached to the leading edges of these structures. The boots are caused to pulsate by compressed air, so that the ice tends to crack and be blown off in the air stream. On rotating surfaces, such as propellers, anti-icing fluid is spread over the structure by the centrifugal force of the rotation, preventing the ice from forming. On larger aircraft, heat is commonly used to prevent ice formation, both on the outside structural surfaces and in the carburetor and other parts of the powerplant. Regardless of the method, however, protective action should be taken before ice begins to form, since a thick layer is difficult—and at times impossible—to remove.

Flying weather—enroute. While enroute from one airport to another, the airplane is subjected to all of the weather elements which occur from altitudes of several hundred feet to the altitude limitation of the aircraft. In light aircraft, the cruising altitude may range upward to 15,000 or 20,000 feet, while in modern subsonic jet transports it will normally be as high as 35,000 to 40,000 feet. With the advent of new supersonic aircraft, cruising altitudes will increase to 70,000 feet or more. This means that light aircraft, and to a large extent subsonic jets as well, fly generally within the troposphere where most of the "weather" (e.g., rain, snow, hail, etc.) occurs. Near the cruising altitude of subsonic jets, and certainly at supersonic altitudes, there is little or no weather in the usual sense, but atmospheric problems of other kinds must be faced.

In general, to an aircraft flying within the troposphere, the primary hazards associated with the weather are similar to those which affect takeoff and landing. However, with the advent of instrument flying, the necessity to have visual contact with the ground enroute becomes less critical. Only to the pilot flying under Visual Flight Rules (VFR) is it legally necessary to be able to see landmarks and other aircraft. For the modern commercial transport, navigation aids and radar in the aircraft provide a means of control, so that clouds and other restrictions to vision become largely annoying inconveniences.

While flying at cruising altitude, aircraft are primarily affected by (1) the speed and direction of the winds aloft, (2) thunderstorms and other small-scale violent weather, (3) hazardous conditions associated with fronts, and (4) potential hazards which may be encountered at very high altitudes proposed for supersonic flight.

Winds aloft affect both the time required for an airplane to reach its destination and the distance it can travel before refueling. If the winds are blowing in the same direction as the plane is moving (tail winds), the travel time is decreased; if they are blowing in a direction opposed to the plane's motion (head winds), the travel time is increased. Thus, since upper-level winds in midlatitudes generally blow from west to east, long-distance flights from the west coast of the United States to the east coast, for example, usually require less time enroute than do flights in the opposite direction.

Since the jet stream often is present at flight altitudes used by long-distance commercial transports, pilots generally try to plan a west-to-east flight along airlanes which are in the path of the eastward moving jet stream. In some cases, the ground speed can be increased by 100 to 200 mph. On the other hand, pilots flying from east-to-west try to avoid the jet stream since such flights can be retarded by a like decrease in ground speed. In the case of over-water flights, the amount of fuel which must be carried is greatly affected by the winds-aloft predictions made by the aviation meteorologist. Depending on whether assisting tail winds or opposing head winds are forecast, the difference in passenger or cargo weight which can be carried may amount to a thousand pounds or more for long distance flights made by large aircraft.

Associated with the jet stream is an elusive and invisible hazard known as "clear air turbulence" (abbreviated CAT by pilots). This phenomenon is caused by the rapid change in wind velocity with distance which takes place near the edge of the jet stream. Like the small eddies which form along the sides of a swiftly moving river, clear air turbulence represents a series of disturbances in the otherwise smooth flow. Figure 9-14 shows the general locations of areas of clear air turbulence with respect to the primary core of the jet stream.

Since clear air turbulence occurs in the clear air at high levels, it cannot be seen. Frequently, it is encountered within only very thin vertical layers of the atmosphere (less than two thousand feet thick) and over relatively short distances. Some experimental observations using very powerful land-based radar have detected the presence of clear air turbulence, but, at present, there is no practical means of observing it from an airplane. However, pilots can be advised of the possible occurrence of clear air turbulence in regions where the wind

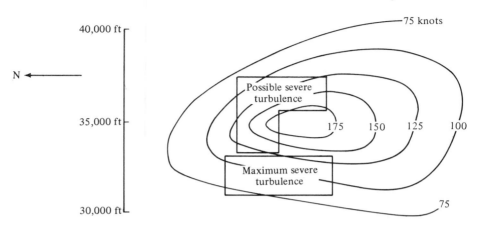

FIG. 9-14.   Usual occurrence of clear air turbulence in the vicinity of the jet stream (Northern Hemisphere).

"shear" (change of wind velocity with distance) is especially great. Such information is derived from the winds-aloft measurements which are made routinely for meteorological purposes.

Thunderstorms can be hazardous to aircraft in flight. Turbulence and hail are perhaps the most serious hazards. Strong vertical currents of air may extend over many thousands of feet in the central core of a thunderstorm. These "drafts," while not necessarily hazardous to the aircraft structure, may force the airplane upward or downward at such speeds that it may be impossible to maintain flight altitude, even with full engine power. Displacements of aircraft of as much as 8000 feet, and vertical speeds up to 200 feet per second (135 mph), have been recorded in the middle of a mature thunderstorm. Superimposed on the larger vertical currents are many small-scale, irregular gusts. These produce the violent turbulent motions which can damage the aircraft and injure the passengers. While draft currents are generally found near the center of a thunderstorm, gusts may be experienced in any portion of the storm and are usually unavoidable by the pilot if he is forced by circumstances to penetrate a thunderstorm.

Hail within a thunderstorm may range in size from as small as a pea to as large as five inches in diameter and may result in severe structural damage to an airplane. The greatest frequency and size of hail is usually found in a mature thunderstorm at levels between 10,000 and 30,000 feet, but it is occasionally encountered even in the clear air outside the thunderstorm itself. Figure 9-15 illustrates what hail can do to an aircraft.

FIG. 9-15. Hail damage to an aircraft. (Courtesy of ESSA.)

Lightning associated with thunderstorms may produce structural damage by perforating the skin of the aircraft and damaging electronic or other equipment. The sudden brilliant flash can also cause temporary blindness. While it is possible for lightning to ignite the gasoline or other fuel, modern aircraft fuel systems are designed to provide protection against this hazard. Nearby lightning may affect the reading of a magnetic compass in an airplane, and the static noise in a radio receiver may be serious even when the lightning is occurring several miles away.

In general, the safest course is to avoid flying through, or even in close proximity to, a thunderstorm. Not only do the hazards represent potentially serious consequences, but occasionally the instability associated with the thunderstorm may result in the even more violent winds of a tornado. In a tornado, the winds may exceed 300 mph in a vortex less than 100 feet in diameter, a combination which may well prove fatal to an aircraft.

Fronts often produce weather hazards to aircraft in flight. The clouds which usually accompany fronts may require flight on instruments, thunderstorms and turbulence are particularly prevalent in frontal zones, and the wind may change abruptly in direction and speed across the frontal boundary. It should be stressed, however, that the weather associated with fronts is not always hazardous and that not all hazardous weather is associated with fronts. The aircraft pilot therefore requires detailed information concerning the characteristics of each individual frontal zone and the weather which accompanies it.

Frontal cloudiness and its influence on the airplane in flight varies greatly with the nature of the front itself. In general, warm-front clouds are usually stratiform types, there is little turbulence, and the primary influence is the restriction to visibility. In fact, pilots frequently find that exceptionally smooth flight conditions are found in the temperature inversion within the warm frontal zone. Cloudiness associated with cold fronts, however, tends to be cumuliform, with considerable turbulence and frequent thunderstorms along the front. Occasionally, also, a "squall line" develops in the warm air ahead of the front (Figure 9-16). These "walls" of clouds may extend to an elevation of 40,000 feet, with individual thunderstorm clouds rising as high as 60,000 to 70,000 feet—up into the stratosphere.

Cold-front type occlusions are usually associated with weather typical of cold fronts, although considerable stratiform cloudiness will

FIG. 9-16. An aerial view of a portion of a squall line. (Courtesy of ESSA.)

be found in the region of the upper warm front which is present at high levels behind the surface cold front (Chapter 6). Warm-front occlusions, on the other hand, are accompanied by weather typical of warm fronts, except that frequent turbulence, cumuliform clouds, and thunderstorms will be found near the upper cold front which precedes the surface warm front. Since it is normal practice to show only the surface occluded front on the usual weather chart, pilots should be particularly careful to inquire concerning the positions of the fronts aloft before taking off. Frequently, in such cases, the most severe weather is encountered at the location of the upper front, some distance from the occluded front at the surface.

An airplane flying through a frontal zone will usually encounter a sharp change in temperature and moisture. In general, the temperature change will be more abrupt at low levels than at high altitudes, and the cold air mass will be drier than the warm air mass. A sharp change in wind direction, and frequently wind speed also, will be observed by an aircraft crossing a front. Since the shift in direction is always cyclonic (counterclockwise in the Northern Hemisphere), the aircraft heading should always be changed toward the right when flying in the Northern Hemisphere in order to maintain the same ground track. This rule applies regardless of the type of front, or the direction the aircraft is flying through it. Where there is a very sharp change in wind direction and speed, there may be, in addition, a problem of maintaining airspeed.

With the advent of new supersonic aircraft flying at high levels in the stratosphere, the influence of thunderstorms, fronts, and other tropospheric hazards should be of no concern except during the ascent and descent of the airplane. Nevertheless, the atmosphere does pose important aircraft design and operating problems. The most familiar of these is the compression wave produced by the airplane pushing the air ahead of it at high speeds and causing the "sonic boom." At present, there appears no way to eliminate this phenomenon when the aircraft exceeds the speed of sound. The high concentration of ozone at or near the flight altitude of supersonic aircraft is also of concern. Since ozone readily oxidizes most substances, corrosion may be a serious problem. Although the development of new metal alloys and synthetic plastics will alleviate this difficulty, more frequent replacement of critical parts may be necessary.

Aviation weather services—observations. Many of the basic meteorological measurements described in Chapter 1 are useful for aviation purposes as well. However, because of the critical requirement for visual contact with the airport runway upon takeoff and landing, a great deal of emphasis is placed by the pilot upon measurements of vertical and horizontal visibility. This has led to the development of special instruments: a *ceilometer* for measuring cloud height, and a *transmissometer* for measuring the horizontal transmissivity of the atmosphere.

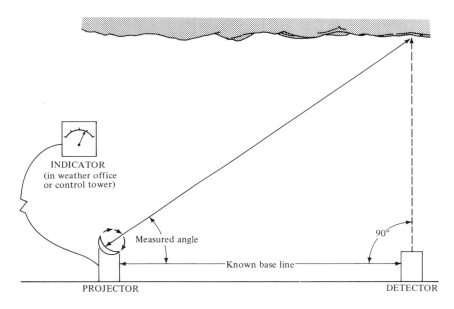

FIG. 9-17. Principle of the rotating-beam ceilometer.

The ceilometer is usually located at the "middle marker" of an airport (about one-half mile from the touchdown point along the path taken by a plane coming in for a landing). The principle of operation is illustrated in Figure 9-17. A projector which contains a high-intensity rotating light source is located a known distance from an "electronic eye" (photocell) detector. As the light beam rotates, the illuminated spot it makes on the base of the cloud is observed by the detector, and the elevation angle of the light beam is measured and transmitted to the indicator. Since the distance from the projector to detector is known, and the detector is oriented vertically, the height of the cloud base can be computed as a simple problem in trigonometry.

The transmissometer system is normally located near the touchdown point of an instrument runway. The principle of its operation is illustrated in Figure 9-18. A projector containing a strong light source

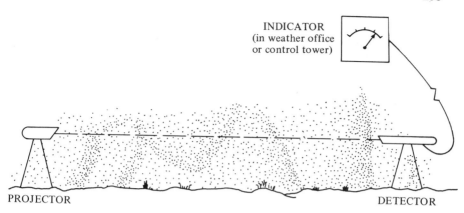

INDICATOR
(in weather office
or control tower)

PROJECTOR

DETECTOR

FIG. 9-18. Principle of the transmissometer.

is aimed horizontally along the direction of the runway toward an electronic eye detector located a known distance away. When there are no obstructions along the path of the light, the detector registers the full illumination produced by the light source. However, if fog, smoke, or other obstructions occur, the light intensity measured by the detector is reduced. The amount of this reduction is calibrated in terms of the average human ability to distinguish objects at various distances, and the result is interpreted as "visibility." However, human visibility depends not only upon the transmissivity of the atmosphere, but also on the brightness of the objects being observed and other factors. In an effort to compensate for some of these effects, the reported visibility is modified by a small computer in the transmissometer system; the modification depends upon the relative brightness of the runway lights used as guidance by the pilot upon landing. This modified measurement is called the "runway visual range."

Another meteorological observation which is provided especially for the pilot is the *altimeter setting*. The need for this measurement arises because the altimeter which is carried in every airplane cockpit to indicate altitude is essentially an aneroid barometer. We have already seen that the pressure above a given point depends upon the weight of the atmosphere, which in turn is dependent largely upon the vertical temperature distribution above that point. Accordingly, in order to calibrate an altimeter in terms of height, it is necessary to specify the nature of the temperature distribution which is to be used. This assumed relationship between pressure and height is the Standard Atmosphere, given in Appendix 8.

Of course, these standard conditions are seldom realized in the real atmosphere, so that the actual altitude is rarely the same as that shown by the altimeter. In general, if the barometric pressure is lower than specified by the Standard Atmosphere, the airplane will be lower than indicated by the altimeter. Similarly, if the temperature is colder than the Standard Atmosphere, the airplane will be lower than indicated by the altimeter. These are therefore dangerous situations, for a pilot may have insufficient altitude to clear a mountain peak if he is lower than anticipated. If the pressure and/or temperature are higher than standard, the airplane is higher than indicated by the altimeter. While this situation will provide sufficient clearance above ground, it may place the pilot at an altitude where other traffic is flying on a different heading; thus, the chance of a collision may be increased.

In order to compensate, at least in part, for these errors, an "altimeter correction" is made by the pilot before takeoff, and routinely throughout his flight if he is flying at altitudes of 18,000 feet or lower. Above 18,000 feet, the effectiveness of the correction is decreased and all altimeters are required to be set to the Standard Atmosphere; this means that, although the height above sea level is not known accurately, all aircraft have the same relative error and therefore have accurate vertical separation.

A more sophisticated device called a *radio altimeter* makes use of the principles of radar to indicate the absolute height of the aircraft above the ground. This is accomplished by emitting a radio signal downward from the aircraft and measuring the time taken for the signal to be reflected from the ground. The amount of time is proportional to the height above ground. The radio altimeter is not subject to the pressure-height errors which afflict the pressure altimeter, but of course the elevation of the ground surface from which the radio beam is reflected must be known. In mountainous country, this may result in considerable uncertainty. Furthermore, the radio altimeter is relatively large and expensive; therefore, it is generally installed only on large commercial aircraft.

Aviation weather services—reports. Observations for aviation purposes are made regularly every hour at all major airports throughout the United States, and at a great many secondary airports as well. In addition, special observations are made more often during critical or rapidly changing weather. These observations are transmitted over special aviation teletype circuits as "aviation weather reports" or "weather sequences." The reports are distributed in a shorthand code form in order to save communications time. Appendix 6 contains a sample of an aviation weather report, and an abbreviated explanation of the code itself.

Aviation weather services—forecasts. Predictions of weather for aviation purposes are made in a number of categories to fit the needs of pilots for takeoff, in-flight, and landing operations. *Terminal forecasts* are prepared for all major airports in the United States and are transmitted over the aviation weather teletype circuit in the same manner as aviation weather reports. They are made for periods up to 24 hours in advance, and include specific information required by pilots for takeoff and landing; e.g., amount and height of clouds, visibility, weather (rain, snow, etc.), obstructions to vision (fog, haze, etc.), surface wind velocity and, when appropriate, remarks concerning other pertinent phenomena.

Area forecasts are predictions of the weather over large regions and are intended for use while the aircraft is in flight. They are transmitted over the aviation teletype circuit in a combination of code format and abbreviated plain language. Data provided in area forecasts include predictions of cloud tops, icing and height of the freezing level, turbulence, and hazardous weather which may be encountered by flights in the area.

Winds-aloft forecasts are predictions of the expected wind direction and speed at various levels for use in flight planning. They are issued in a coded form and cover a period of 12 hours.

A number of other aviation forecasts and warnings are issued, generally for the purpose of drawing attention of pilots to impending weather developments that may cause hazardous flying conditions. Such specialized forecasts are generally transmitted in plain language, or with easily deciphered abbreviations. An example of each type of routine aviation forecast is given in Appendix 6.

9-4. Industry

As in many other forms of human endeavor, weather constitutes a significant influence on the efficiency of industrial activities. Such influences include not only the more catastrophic storms—tornadoes, hurricanes, or blizzards—but even the more moderate forms of "bad" weather such as a few inches of snow, an all-day rain, or several days of freezing temperatures. In many cases, these more moderate weather phenomena can affect an industrial operation more seriously than a spectacular storm. For example, even hurricane winds will seldom damage a well-constructed building or other structure, but construction of an industrial plant may be delayed seriously by rain while the building site is being excavated, by moderate winds during the erection of steel girders, or by subfreezing temperatures when concrete is being poured. In this section, we shall consider a few of the typical weather problems which are faced by the growing, worldwide industrial complex.

Planning of plant location. Because the construction of a new industrial plant usually represents a major capital investment which must be expected to last for many years, the initial planning of its location and design is an extremely important undertaking. In some cases, the nature of the operation will be vitally affected by the general weather conditions or climate of the area in which it is located. For example, a manufacturing concern which is engaged in the fabrication of delicate electronic equipment may require expensive air-conditioning facilities if it is located in a region where the atmospheric temperature or humidity vary beyond the tolerance limits of the manufactured product. Another plant may find that bulky raw material which it uses in a manufacturing process must be stored out of doors because it would be too expensive to construct a warehouse for storage purposes. If the material is affected by certain weather conditions (e.g., freezing, strong winds) it is important that the plant be located in a region where such weather does not occur—or occurs only rarely.

Perhaps one of the most important problems which must be considered in planning a modern industrial plant is the amount and nature of the smoke, fumes, and other pollutants which will be emitted into the atmosphere. Here, the problem may involve not only the normal (and perhaps relatively harmless) emissions which take place during the operation of the plant, but in some activities, certain catastrophic consequences which may result from an accidental discharge of poisonous chemicals or radioactive material into the atmosphere. Where such problems arise, it is important that the plant not be located in a region where the pollutants will remain concentrated in the atmosphere, or where the prevailing winds may be expected to cause them to drift into a populated area.

As we have seen in Chapter 4, when the atmosphere is stable there tends to be little dispersion of contaminants and, in particular, a temperature inversion at low levels will be most favorable for the concentration of "smog" near the ground. Such conditions typically occur as a result of subsidence in a high-pressure system, and/or as a consequence of nocturnal radiation during clear, calm nights. Where such atmospheric conditions are prevalent (as, for example, in many areas of the semipermanent, high-pressure belt of the horse latitudes) the location of large industrial centers clearly appears unwise. Considering the world as a whole, the atmosphere provides the most efficient pollution-dispersing mechanism in those regions of generally rising air motions—the tropics and midlatitude storm belt associated with the polar front.

Other atmospheric phenomena which should be considered in planning the location of an industrial plant include the prevalence of ice and snow, the frequency of critically high or low temperatures, the amount and annual distribution of precipitation, and the likelihood of damaging winds. Not all of these weather elements will be significant for all operations and, of course, other non-meteorological factors must be considered; e.g., availability of raw materials, labor supply, and the like.

Plant construction. Weather influences on the efficiency of the construction industry cover a wide range of elements. Beginning with the surveying and clearing of the construction site, through the completion of the final structure, unfavorable weather may cause lost or unproductive work days, equipment may be forced to remain idle, and construction material may be damaged or destroyed. Some idea of the more significant weather factors which may affect the construction, not only of industrial plants, but also of such other structures as houses, bridges, office buildings, etc., is given in Table 9-7.

Because of the critical nature of many of the operations listed in Table 9-7, as well as the small range of weather conditions within which some operations can be carried out, many aspects of construction must be performed only at certain times of year. Thus in northern areas of the United States, outdoor concrete pouring and other activities must usually be completed during the summer months. Similarly, most construction during the hot summers of the desert regions is hampered by temperatures in excess of 90°F. In some cases, of course, work may be carried on but less efficiently.

Plant operation. Many manufacturing and other industrial operations are significantly affected by weather conditions. Severe weather, such as heavy snow, torrential rains, or the like, may disrupt the flow of traffic so that employees may find it difficult or impossible to get to work. Similar weather may also interfere with the orderly flow of raw material, or the efficient distribution of manufactured products. Some assistance in alleviating these disruptions can be provided by the meteorologist in the form of weather predictions—adequate warnings

TABLE 9-7

Critical Limits of Weather Elements Having Significant Influence on Construction Operations (Abridged from Russo, et al.)

Operation	Rain	Snow and sleet	Freezing rain	Low temp. (°F)	High temp. (°F)	High wind (mph)	Dense fog	Ground freeze	Drying conditions	Temperature inversion
Surveying	L	L	L	−10–0	90	25	X	—	—	—
Site clearing & grading	M	M	L	0–32	90	15–25	X	X	X	X
Site excavation	M	M	L	20–32	90	35	X	X	X	X
Pouring concrete	M	L	L	32	90	35	—	X	X	—
Erecting structural steel	L	L	L	10	90	10–15	X	—	—	—
Exterior carpentry	L	L	L	−10–0	90	15	—	—	—	—
Exterior masonry	L	L	L	32	90	20	—	X	X	—
Roofing	L	L	L	45	90	10–20	—	—	X	—
Exterior painting	L	L	L	45–50	90	15	X	—	X	—
Landscaping	M	L	L	20–32	90	15	X	X	X	—

L = light amount.
M = moderate amount.
X = operation affected but amount undeterminable.

of severe weather may permit plant managers to reschedule production and distribution so that the difficulties are minimized.

Other, less dramatic weather conditions may also require that special production schedules be developed. If air pollution poses a potential problem in the plant operation, certain activities may have to be curtailed at certain times of day, or in certain seasons of the year when local temperature inversions are more common.

Figure 9-19 shows the frequency of temperature inversions at low levels (below 1500 feet) at Atlanta, Georgia during the morning (10:00 A.M.) and evening (10:00 P.M.) throughout the year. Note that the percentage frequency of such inversions, and thus the air-pollution potential, remains high at night all year, but daytime solar heating almost eliminates the inversion during the summer days. Since wind speeds tend to be somewhat higher during the summer days in this area, ventilation of possible pollution is greater at that time of year as well.

Other weather elements which represent sources of concern in the operation of an industrial plant include (a) the occurrence of temperatures below the freezing point of liquids carried in outside pipe

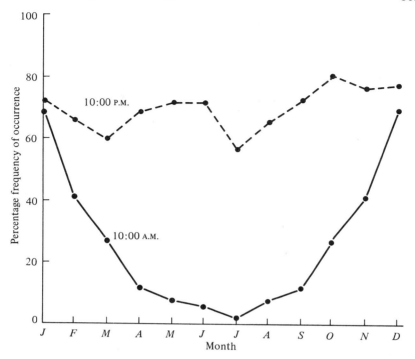

FIG. 9-19. Per cent frequency of temperature inversions below 1500 ft at Atlanta, Georgia.

lines; (b) winds in excess of speeds which would endanger maintenance activities on scaffolding or roofs of buildings; (c) wind-driven tides in a river or estuary which may flood low-lying areas of the plant; (d) high temperatures and low humidities which may produce a potential fire hazard; and (e) certain otherwise uncritical elements which are of concern because of the special nature of the product being manufactured.

In order to deal effectively with problems of weather, an increasing number of industrial organizations employ professional meteorologists whose interests and training include both meteorology and the speciality of the company. Thus, there has arisen—particularly in the United States—a group of highly trained professionals with such dual interests as meteorology and engineering; meteorology and chemistry; meteorology and geology; and many others. Private consulting firms, not necessarily associated with a particular industrial organization, are also available to provide meteorological advice, climatological analyses, and other specialized services not generally available from government weather agencies.

9-5. Marine Activities

The influence of the weather has always been a major factor in the lives of those who make their living from the sea. Before the advent of steam engines, sailing ships were dependent upon the winds as a source of power, and were usually required to travel one course from the home port to a destination and a different route to return. Thus, for example, the northeast trade winds were used in sailing from Europe to the New World, while the prevailing westerlies provided the winds for the return trip. This meant that it was usually not possible to travel the shortest distance between ports and, because adequate winds were necessary to sail at all, ships frequently encountered gale winds in severe storms and suffered substantial damage or were lost.

Even with the development of powered ships, weather still remains a problem. In 1954, over 30,000 medium and large ships were carrying the ocean commerce of the world. Of these ships, more than 1000 suffered heavy weather damage enroute between ports, about the same number were stranded for varying lengths of time by strong winds, and 20 foundered and were a complete loss. In addition, a large amount of physical damage to cargo resulted from rough seas whipped up by the winds, and by sea water or condensed moisture in the hold.

In order to alleviate these difficulties, the meteorological services of the world have for many years provided forecasts and warnings to all mariners concerning the development and movements of severe storms over the oceans. In fact, it was as a consequence of the maritime tragedies of the nineteenth century that meteorological services in many countries were first established. In recent years, however, these warnings of severe weather have been supplemented by more detailed advice which permits the ship captain not only to avoid major storms, but also to select a course which will ensure the fastest and most comfortable voyage for passengers and crew. Clearly, such a course will also result in speedy delivery and minimum damage to cargo.

Least-time tracks. The shortest distance for a vessel traveling between two points on the earth is known as a "great-circle" course. However, this may not always be the course involving the shortest travel time, since the winds and rough seas along a great-circle course may delay a ship so much that another ship sailing another course with favorable weather will make better time. A course which takes account of the favorable weather and other factors and results in the minimum travel time is known as a "least-time track."

The important factors to be considered in determining a least-time track are the winds, the ocean waves, and the ocean currents. However, since the day-to-day winds produce the ocean waves, and the average winds over the oceans are related closely to the ocean currents, the selection of a least-time track for a particular voyage is primarily the task of the meteorologist. The process usually involves three steps:

1. Least-time tracks based upon seasonal changes in the wind patterns can be obtained from climatological data which summarizes the average wind-, wave-, and ocean-current patterns for each season of the year. The result of such a summary in the Pacific Ocean, for example, would suggest that ships sailing between California and Japan should follow close to a great-circle route during the summer, reaching as far north as latitude 47 degrees in mid-Pacific. During the winter, a more southerly course should be followed in order to avoid stormy weather; this would add about 800 miles to the voyage, but the time enroute would be shorter than if the summer route were followed.

2. Long-range weather forecasts of from five days to two weeks may improve on the seasonal routes if the forecast during a particular period indicates that little or no stormy weather is in prospect for the great-circle track. For example, a long-range forecast of good weather over the north Pacific during the winter would permit a ship to sail from California over the usual summer route and, thus, the additional 800 miles of travel could be avoided. It should be noted that such voyages may be somewhat of a gamble because of the current relatively low accuracy of long-range predictions; however, as meteorological science progresses, such forecasts may be expected to improve.

3. Short-range forecasts up to 48 hours in advance may also be used. In this case, the course would be planned from day to day, so that the prospective track may be changed throughout the voyage. Because short-period forecasts are usually quite accurate, this procedure can be used very effectively in modifying a basic course which was based upon climatological data and long-range weather forecasts.

The actual voyage will normally be the result of all three procedures. The seasonal weather and ocean patterns are first considered and are then modified on the basis of the long- and short-range predictions. In connection with the latter, for example, an early departure may permit a ship to make use of favorable winds accompanying a

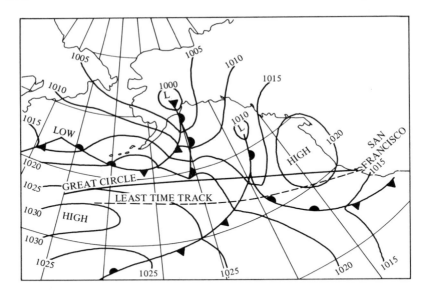

FIG. 9-20. Least-time track forecast for voyage from San Francisco to Yokohama (Sept., 1954).

slow-moving, high-pressure system, or a delay may allow an intense storm to move beyond the range of the prospective track.

Figure 9-20 shows an example of a weather map on which has been drawn the great-circle (shortest-distance) course, compared with a computed least-time track. Note that, since the storms (low-pressure centers) are evidently moving eastward along northern latitudes, the least-time track would take the ship south of the great-circle course through the light winds of the high-pressure system.

Other marine activities. In commercial and sport fishing, meteorology plays a significant role. This not only applies to the need for providing weather forecasts and warnings of severe storms, but also to a growing recognition of the relationships that exist between ocean conditions and the concentration of certain species of fish. While the sun is the major source of energy which raises the temperature of the sea, the mixed layer near the surface in which the majority of food fish live is the result of turbulence produced primarily by the winds; furthermore, the movement of warm and cold water masses is affected by the winds.

The most common food fish of the temperate zone; e.g., the cod family and the herring, exist in waters with a temperature range of between about 5°C (41°F) and 15°C (59°F). When water temperatures are colder than about 5°C, the number and variety of food fish decreases. In the tropics, where water temperatures are higher than 15°C, the variety of all forms of life increases, but the number of any one kind of fish decreases. However, some of the favorite fish of the sport fisherman are found in warm, tropical waters. The bluefin tuna, for example, is most abundant in the Caribbean Sea and follows the warm Gulf Stream northward as far as Maine and Nova Scotia during the early summer months. Within recent years, weather forecasts have been made especially to assist fishermen in tracking the ocean currents associated with optimum temperature conditions for food and sport fishing.

Off-shore oil drilling and other mining activities on the ocean floor have increased rapidly during the past few years as man has attempted to replenish his ever-dwindling store of natural resources. These complex operations are extremely sensitive to environmental changes in both the ocean and the atmosphere. In the Gulf of Mexico, for example, oil-drilling rigs on the continental shelf must be evacuated during the approach of a hurricane because of danger to the lives of the crew members who operate the drilling apparatus. The construction of piers, wharves, and other man-made structures along the coast and in harbors are additional examples of fixed installations which are subject to weather hazards.

Recreational boating, swimming, skin diving, and surfing have also increased rapidly during the past decade. In 1964, there were nearly 8 million registered owners of small craft of various kinds in the United States. During that year, property losses among small-boat owners amounted to nearly half a million dollars and 216 lives were lost, primarily as a result of accidents associated with adverse weather conditions. In many cases, these accidents were aggravated by ignorance of the warnings of severe weather provided by the United States Weather Bureau. These warnings are issued not only for ships on the high seas, but also for small craft sailing along the shore and on inland waters. They are broadcast by radio over both government and commercial radio stations, and visual signals are displayed in a prominent location at many port offices, yacht harbors, and lifeguard stations. Figure 9-21 illustrates the signals displayed to indicate the range of wind speeds and sea conditions which may be dangerous to both large and small vessels.

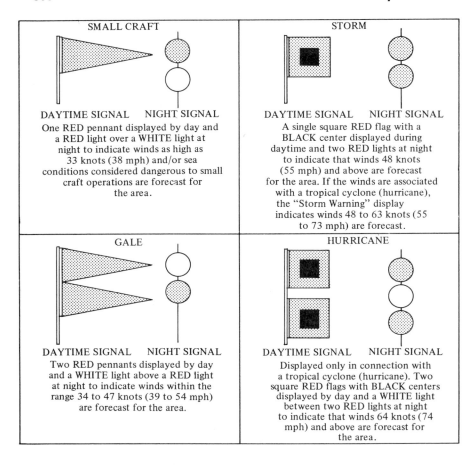

FIG. 9-21. Marine display signals. (Shading indicates red flag or light; solid area on flag is black; open circle indicates white light.)

The following are examples of weather rules of safe boating which should always be observed:

Before setting out:
- Obtain the latest weather reports and forecasts from radio broadcasts, or by telephoning the nearest office of the United States Weather Bureau.
- If warnings are displayed, decide whether your boat can be navigated safely at wind speeds or in sea conditions forecast for the area.

While afloat:

- Always have a radio receiver aboard, and keep a continuous check on the latest forecasts and warnings. Always watch out for:
  - (a) the approach of dark, threatening clouds which may indicate an impending squall or thunderstorm;
  - (b) any steady increase in wind speed or wave height, or both;
  - (c) any increase in wind velocity opposite in direction to a strong tidal current; this may produce a dangerous "rip tide" capable of overturning a small boat.

9-6. Military

From the time of man's first conflicts, weather has had a significant influence in determining the outcome of many critical battles. Classic examples include the Spanish Armada of 1588 which was destroyed, not primarily by the British fleet, but by gale winds and accompanying seas which wrecked many of the ships on the coasts of Scotland and Ireland. The invasion of Russia by Napoleon's armies in 1812 was doomed by the blizzard conditions which were encountered on the return journey from Moscow. And an unexpected storm in the Black Sea during the Crimean War in 1854 resulted in the sinking of more than 30 ships of the French and British fleets and the subsequent failure of the campaign.

This last disaster provided the major stimulus for the establishment of the first weather forecast services. The French government commissioned Jean LeVerrier, director of the Paris Astronomical Observatory, to study the meteorological conditions associated with the Black Sea Storm.* By assembling the weather reports obtained from other European observatories, LeVerrier was able to chart the development and progress of the storm, and to show that the newly invented telegraph could be used to transmit such data quickly enough to be used for forecasting purposes. From this beginning, a daily weather-observing and reporting network was organized, subsequently resulting in the establishment of a World Meteorological Organization which today serves as a means of coordinating the exchange of weather information among weather services of all countries of the globe.

Modern military operations have become so complex that the effects of weather are even more critical than in previous times. During World

---

* For other facts about the historical development of meteorology, see Appendix 1, "A Brief Chronology of Atmospheric Science."

War II, the timing of the invasion of the Normandy beaches by the Allied Powers was dependent largely upon forecasts of weather and sea conditions. The landing on the beaches could not be accomplished during periods of high seas or strong winds. Supporting aircraft needed adequate ceiling and visibility to carry out their missions, while tanks and other heavy equipment could become mired on the beaches and in the fields if heavy rains occurred during the operation. In actual fact, the landings were made during a period of predicted marginal weather. This resulted in a somewhat hazardous operation, but caught the opposing forces by surprise since they had not expected that the invasion would be attempted under such conditions.

Today's military applications of meteorology include a great many aspects beyond even those of World War II. Refueling of long-range aircraft in flight is a routine operation which requires accurate predictions of cloudless skies and smooth flying conditions. The testing and operation of ballistic missiles is dependent upon a knowledge of the winds, cloudiness, temperature and other factors for launching and re-entry of such vehicles from outer space. Even ground troops and conventional artillery are now supported by weather observations and forecasts which provide for more efficient and effective operations.

9-7. Decision Making

One important characteristic of meteorological information is the inherent uncertainty in both observations of the current weather and predictions of its future state. This difficulty arises for a number of reasons. In the first place, the networks of surface stations, radiosondes, satellites, etc., which are used to observe the atmosphere, provide only a crude measure of its small-scale variations. This means that the uncertainty begins, in fact, with the lack of precision with which the initial state of the atmosphere can be determined. Other restrictions arise because of a technological inability to obtain an exact formulation or solution of the prediction problem, and because of ignorance concerning certain influences, internal or external to the atmosphere, which may be important.

As science advances, these deficiencies will gradually be overcome. For the forseeable future, however, the uncertainty with which they are associated will be of concern to virtually all of those who must make practical decisions concerning an activity which is sensitive to the weather. The question then arises: Recognizing that some lack of precision is inherent in weather information, and that perfect decisions based on that information are not possible, how may a user of the weather forecast insure that his decisions at least will result in the greatest economic profit or the least loss?

Simple decision theory. The following discussion is based on the fundamental principles of modern decision theory. It represents a somewhat simplified, but nevertheless perfectly general approach to the above question.

Consider the case of a prospective user of a weather forecast faced with the problem of deciding whether or not to take protective measures against a certain adverse weather element. The user may be, for example, a farmer, a business man, or an airline operator, and the weather element may be snow, rain, low ceiling, or poor visibility, or the like. In general, the user should take protective measures against the occurrence of adverse weather if, in the long run, an economic or other gain will be realized; otherwise, no protective measures should be taken. In order to derive a criterion for making this decision, the following terms are defined:

f_w = Frequency (number of cases) of adverse weather.

f_{nw} = Frequency (number of cases) of favorable (non-adverse) weather.

N = $f_w + f_{nw}$ = total frequency (adverse + non-adverse weather cases).

G_p = Total expected gain for N occasions if protective measures are taken.

G_{np} = Total expected gain for N occasions if protective measures are not taken.

C = Cost of protection on each occasion that protective measures are taken.

L = Loss suffered on each occasion that adverse weather occurs and no protective measures have been taken.

T = Average net profit for each occasion, exclusive of the cost of protective measures, or the loss which may have been suffered.

If, now, protective measures are taken on N occasions, the total gain will be the profit minus the cost of protection, both multiplied by the number of occasions. Or, in symbolic form,

$$G_p = (T - C) N \qquad (9\text{-}2)$$

If protective measures are not taken, the gain can be divided into two parts—the reduced profit during cases when adverse weather occurred and the profit for the cases of no adverse weather; in symbolic form,

$$G_{np} = (T - L)f_w + Tf_{nw} \qquad (9\text{-}3)$$

Since the user should take protective measures only if the gain by so doing will be greater than the gain by taking no protective measures, protection should be provided when $G_p > G_{np}$ or, from Equations (9-2) and (9-3),

$$(T - C)N > (T - L)f_w + Tf_{nw} \qquad (9\text{-}4)$$

which reduces to simply

$$f_w/N > C/L \qquad (9\text{-}5)$$

The left-hand side of this expression defines "P," the probability that adverse weather will occur. Note that P is the "relative frequency" of adverse weather, or the number of occurrences of adverse weather divided by the total number of cases. In a similar manner, it may be shown that protective measures should not be taken if $P < C/L$, and either course may be followed if $P = C/L$. Thus, a criterion for making a decision to protect, or not protect, against adverse weather may be expressed,

$$P \begin{Bmatrix} > \\ = \\ < \end{Bmatrix} C/L \qquad \begin{matrix} \text{Protect} \\ \text{Either course} \\ \text{Do not protect} \end{matrix} \qquad (9\text{-}6)$$

The value $P = C/L$ thus represents a critical ratio, above which protection should be provided, and below which it should not. It is interesting to observe that for C, L, and T, as defined here, the profit drops out and need not be considered in the decision. Alternative, but generally more complex expressions may be derived by defining these terms in a different manner.

Now it is clear that the user of the weather information must provide the numerical values of C and L. They are the costs and losses associated with the weather-dependent portions of the operation. The value of P, the probability of adverse weather, must be provided by the meteorologist. This may be done by an analysis of past experience. For example, with a particular set of initial surface- and upper-air conditions, the meteorologist knows, or can estimate quite well, the relative frequency with which adverse weather has occurred under similar conditions in the past. If a given weather map has resulted in only a relatively few occurrences of adverse weather in the past, the probability of adverse weather will be low. If adverse weather has occurred frequently under these conditions in the past, the probability of adverse weather will be high.

An operational problem. To illustrate the principle involved in the preceding discussion, consider the following example. A contracting firm has a large amount of equipment which would be damaged if it were left outdoors during periods when the temperature fell below freezing. Accordingly, it is the custom for the equipment to be brought inside every night during the winter months, an operation which costs the company about 300 dollars in wages and material each time this is done. However, if the equipment were left outdoors when temperatures fell below freezing, the loss in damaged machinery would amount to 2000 dollars.

The superintendent of the plant feels that a portion of the daily cost of protection might be saved if the equipment were left unprotected when the weather forecast called for temperatures above freezing, and only brought inside when below-freezing weather was predicted. He therefore obtains from the weather office a list of the daily minimum temperatures, predicted and observed, during the previous winter. Summarizing these under the assumption that the equipment would be brought inside whenever freezing weather was predicted, and left outside on other days, he obtains the following table:

TABLE 9-8

Analysis of Decisions Based on Conventional Weather Forecasts
(Figures in Body of Table Are Frequency of Occurrence)

	Course of action		
Observed weather	*Protection*	*No protection*	*Total*
Freezing or below	51	6	57
Above freezing	6	27	33
Total	57	33	90

Noting that protective measures would have been taken on 57 occasions, and that no protection would have been provided but freezing was observed on 6 occasions, the superintendent computes the total expense of this operation to be:

$$\text{Expense using conventional forecasts} = 57 \times \$300 + 6 \times \$2000$$
$$= \$29,100$$

However, if the forecasts were not considered, but the current practice of taking protective measures every day was continued, the total expense would be:

$$\text{Expense without forecasts} = 90 \times \$300 = \$27,000$$

It appears at first glance, therefore, that the forecasts are not good enough for the purposes of this operation. Before coming to a final conclusion, however, the superintendent decides to visit the weather office and ask whether any way can be found to alleviate this difficulty. The meteorologist to whom he presents these facts replies that such a way can indeed be found, and describes the principles outlined in the preceding section of this chapter. Noting that the company's cost of protecting their equipment is, in this case, 300 dollars, and the loss if it is not protected is 2000 dollars, the meteorologist computes the ratio $C/L = 300/2000 = 0.15$. Thus, the optimum decision criterion (Equation 9-6) for the operation can be stated

$$P \begin{Bmatrix} > \\ = \\ < \end{Bmatrix} 0.15 \quad \begin{matrix} \text{Protect} \\ \text{Either course} \\ \text{Do not protect} \end{matrix} \quad (9\text{-}7)$$

The meteorologist then presents a table which he has computed by tabulating, for an independent two-year period, the number of times that 32 degrees or below was observed following the daily temperature predictions issued by the weather office. Then, by classifying these data according to the predicted temperatures, it is possible to obtain the relative frequency, or "probability" of occurrence of freezing weather following any predicted temperature, as shown in Table 9-9.

TABLE 9-9

Probability of Freezing Weather from Forecast Verification Data

Predicted temperature	For each predicted temperature		
	Total observed frequency (a)	*Frequency of observed temperature* $\leq 32°$ (b)	*Probability of temperature* $\leq 32°$ (b/a)
≤ 14	5	5	1.00
15–19	12	12	1.00
20–24	24	23	0.96
25–29	36	31	0.86
30–34	44	28	0.64
35–39	23	10	0.44
40–44	20	3	0.15
45–49	12	1	0.08
≥ 50	5	0	0.0

An inspection of these data reveals that the critical probability (0.15) for this example is exceeded whenever temperatures in the class interval 40–44 degrees or lower are predicted. Assuming that, for practical purposes, the decision to protect can be made at the midpoint of this class interval, a second set of verifications is prepared wherein protective measures are taken whenever 42 degrees or below is predicted and none taken when temperatures above that value are forecast. The result of this experiment is given in Table 9-10.

TABLE 9-10
Analysis of Decisions Based on Forecast Verification Probabilities

Observed weather	Course of action		
	Protection	No protection	Total
Freezing or below	56	1	57
Above freezing	22	11	33
Total	78	12	90

From Table 9-10, the expense of carrying on the operation is computed to be:

$$\text{Expense using verification probability} = 78 \times \$300 + 1 \times \$2000$$
$$= \$25{,}400,$$

or 1600 dollars less than the cost of protecting the equipment every day, and 3700 dollars less than basing the operational decision on the forecasts as originally issued.

At this point, the meteorologist suggests that although this procedure is indeed useful, a better method is to ask the forecaster to provide the forecast probability each day, and that his staff has already made such estimates for each day of the previous winter. Using these estimates, and assuming that protective measures would be taken whenever the probability of freezing weather exceeded 0.15, Table 9-11 is obtained.

TABLE 9-11
Analysis of Decisions Based on Forecaster's Estimated Probabilities

Observed weather	Course of action		
	Protection	No protection	Total
Freezing or below	56	1	57
Above freezing	14	19	33
Total	70	20	90

From Table 9-11, the expense of carrying on the operation is computed to be:

Expense using forecaster's probability = 70 × $300 + 1 × $2000
= $23,000,

or 2400 dollars less than the previous procedure and a saving of 4000 dollars over the company's current operation which involved taking protective measures every day.

This example illustrates an important concept in the efficient use of weather information. For the expenditure of a few hours time on the part of the meteorologist and the businessman working together, a saving of 4000 dollars over the three-month period of this example could be achieved. The example itself is somewhat oversimplified; yet it illustrates an important principle concerning the meteorologist's service to the business, industrial, and agricultural communities. It shows how the meteorological profession can help decision makers eliminate the "guess work" in carrying out weather-dependent operations. Furthermore, it does not depend upon waiting for a scientific improvement in weather forecasting, although such an improvement would increase the value of the forecast still further. Its purpose is to insure that the prevailing state of the science is used more efficiently.

Public probability forecasts. For several years, meteorologists have been carrying out experiments to provide the basic information for making the best use of weather forecasts. In some cities, and in a number of agricultural areas, the weather forecast includes the probability of rain occurrence. An example of the kind of prediction issued under this experimental program would read: "Cloudy today; rain tonight; clearing tomorrow. Rain probability today 40 per cent; tonight 80 per cent; tomorrow 30 per cent." It will be observed that the first portion of the prediction is presented in the usual categorical form for those who wish to follow the mentally less-fatiguing policy of letting the forecaster make their decisions for them. For those who wish to take advantage of the probability forecast, however, such information is presented at the end of the prediction.

For the time being, these forecasts are confined to weather elements which can easily be divided into two classes—such as rain and no-rain, snow and no-snow, above freezing and below freezing. As the meteorologist and the user of weather information gain more experience in the making of better decisions, more complex variations of this kind of forecast will undoubtedly be forthcoming.

PROBLEMS

Problems marked with an asterisk (°) are the most challenging.

1. Considering the effect of temperature and precipitation on plant growth (Figure 9-1) and the general circulation of the atmosphere (Figure 5-1), in what general areas of the world would you expect to find (a) grasslands, (b) forests, and (c) deserts? Compare your answer to the classification of climates of Figure 7-13.

2. A farmer plants an early pea crop in Indiana on April 1. If it is a normal year in which the cumulative heat units above a base temperature of 40°F are 90 on April 15, 270 on May 1, 530 on May 15, 865 on June 1, and 1280 on June 15, about what date may he expect to harvest his crop?

3. On a clear, calm night the temperature in an orchard is 28°F at the ground and increases with height (temperature inversion) at a rate of 10°F per 100 feet. If the orchard is heated so that the surface temperature is 35°F, how high will the heated air rise? What will be the temperature at that height?

4. When trees are cut in the forest for lumber, many of the smaller branches and other debris (called "slash") must be burned in order to provide room for new vegetation. Considering the potential fire hazard during this operation, what general weather conditions should prevail when the slash is burned?

5. During the combustion of petroleum fuels in an engine, water vapor is generally one of the gases emitted from the exhaust. How does this explain the formation of "contrails"—the long white streamers frequently observed following the flight of an airplane at very high levels?

°6. The pressure altimeter in an airplane in flight reads 16,400 feet (5000 meters) and the temperature outside the aircraft is −25°C. The pilot checks the "altimeter setting" by radio from the control tower at an airport directly below him and finds that it reads 29.92 inches. Is the airplane likely to be lower, higher, or at the same altitude as indicated by its altimeter? (Hint: See Appendix 8 for the temperature of the Standard Atmosphere at 5000 meters.)

7. On two different days you make a cross-country flight to a destination which is due east of your starting point. The flights are made at the same altitude and air-speed, but on the first day there is no wind at flight level, while on the second there is a moderate north wind. Will your flight time be shorter, longer, or the same on the second day as on the first?

8. Considering atmospheric factors alone, which of the following areas would be preferred for establishing an industrial plant which emitted noxious fumes during its operation: Southern California or Florida? The Scandinavian Peninsula or the Iberian Peninsula?

9. Suppose you are competing in a sailing race from San Francisco to Honolulu during the month of July. Considering the normal winds (Figure 5-2), would you recommend sailing a great-circle course, a track to the north of that course, or a track to the south of it?

10. A young lady says she would carry an umbrella five times even if it did not rain, rather than get her new hat wet once. If the weather forecast says that the probability of rain is 30 per cent, should she carry an umbrella? Why?

°11. A transportation company is shipping some merchandise to another city in uncovered trucks. If rain occurs, the merchandise will suffer damage of 1000 dollars, but the labor and material required to provide a protective covering will cost 150 dollars. The company manager visits the weather office and obtains a list of the conventional forecasts, probability forecasts, and the verification data for a 10-day period as follows:

| | Conventional | Rainfall | |
Day	forecast	Probability	Verification
1	No rain	0.10	No rain
2	No rain	0.40	Rain
3	No rain	0.05	No rain
4	No rain	0.10	No rain
5	Rain	0.60	No rain
6	Rain	0.90	Rain
7	No rain	0.10	No rain
8	No rain	0.20	No rain
9	No rain	0.30	No rain
10	No rain	0.10	No rain

(a) Using the conventional forecasts, what would be the total weather expense for the operation during the 10-day period?

(b) At what probability level should protective covering be provided in order to make the most efficient decisions?

(c) If decisions were made at this probability level and the probability forecasts were used, what would be the total expense for the operation during the 10-day period?

(d) Since only 10 days were used in this example, do you think one could draw any general conclusions from the results? If not, what should be done in order to be reasonably sure of valid general conclusions?

*12. A nurseryman is engaged in protecting his delicate plants from frost. It costs him 200 dollars in material and wages to protect his plants on an average cold night, but he would suffer about 2000 dollars in damages if, in freezing weather, he had not carried out protective measures. He obtains the following data from an evaluation of the weather forecast:

Predicted temperature	Total observed frequency	Frequency of temperature ≤32°
26 or less	4	4
27–29	5	4
30–32	10	7
33–35	15	6
36–38	16	4
39–41	20	2
42 or more	30	1
Totals	100	28

(a) For this sample of 100 days, what would be the total expense of the operation if the nurseryman protected his plants only when a temperature of 32° or lower was predicted?
(b) At what probability level should he protect his plants?
(c) About what predicted temperature would this be?
(d) If he protected his plants by using this latter temperature, what would be his expense for the operation?
(e) If he protected every day, what would be his total expense?
(f) If he did not protect at all, what would be his total expense?

*13. A frequent question asked by the public is: How accurate are weather forecasts? To answer this question, it is not unusual to reply that the forecasts are "R" per cent right, a term which is defined in the following way. If N forecasts of a weather event "W" are placed in a table similar to Table 9-7, the general form of the data would be:

		Forecast		
		W	No W	Total
Observed	W	a	b	$a + b$
	No W	c	d	$c + d$
	Total	$a + c$	$b + d$	$N = a + b + c + d$

and the percentage of right, or correct, forecasts (R) is given by

$$R = \frac{a + d}{N}$$

(a) If "protection" is considered equivalent to a forecast of "W," and "no protection" equivalent to "No W," what is the value of R for the data in Table 9-8?

(b) Using the same considerations, what is the value of R for the data in Table 9-10 and Table 9-11?

(c) Considering the values of R for these three tables, and the values of the corresponding expenses, what do you conclude about the adequacy of the "per cent right" as a measure of the operational usefulness of a weather forecast?

10

Modification of Weather and Climate

10-1. The Magnitude of the Problem

Man has been affecting his atmospheric environment since he first lit fires and moved into caves. Even outside of his shelter, man has been modifying his weather, both intentionally and unintentionally, for a long time. He has done this by changing the character of the earth's surface, by altering the natural constituents of the air, and even by adding heat to the air. But at least until fairly recently, man's influence has been felt only on an extremely small scale.

The great variability of climate over all scales of time during the history of the earth is well established, even though the causes are not as yet settled (Chapter

7). Since climate has changed markedly due to natural causes, might man not be able to induce such changes also? In this chapter we shall discuss the possibilities of man's intervention in natural atmospheric processes. Much of the discussion will be highly speculative, since only a few of the many proposed schemes for modifying the weather have been carried out. However, one can generally rule out some suggestions as impractical and point out the uncertainties in others. In describing each technique for weather modification, we shall draw on our knowledge of atmospheric behavior described in earlier chapters.

The various scales of weather patterns (Chapter 5) must be kept in mind when one considers the feasibility of any particular scheme for deliberate weather modification. On a large or even a medium scale, we cannot hope, at least in the foreseeable future, to effect changes by applications of energy matching those of natural atmospheric processes. For example, it would certainly be difficult to create winds on any but the smallest of scales that would equal those of nature. This is illustrated by the data of Table 10-1; to match the

TABLE 10-1

Approximate Kinetic Energy Contained in Various Atmospheric Circulations and the Time Required to Produce the Equivalent Amount of Electric Energy in the United States

Atmospheric circulations	Kinetic energy (kw-hr)	U.S. production time
Tornado	10^7	5 minutes
Small thunderstorm	10^8	50 minutes
Large thunderstorm	10^9	10 hours
Tropical cyclone (hurricane)	10^{11}	5 weeks
Extratropical cyclone	10^{12}	50 weeks
General circulation of Northern Hemisphere	10^{14}	100 years

energy of the air motion of a single small thunderstorm, we would need all of the United States' electric power for almost an hour (assuming that we could concentrate all this power in a small area). So fans are of little use except within buildings and over small areas (e.g., over orchards where fans may be used to stir up the air enough to prevent shallow radiation inversions from intensifying). But the impracticality of ventilating an entire city to blow away a pall of pollution, as has sometimes been suggested, can be easily demonstrated by computing the energy required to move such a mass of air. Thus, to change the air to a depth of 1 km over a 20 km × 20 km city once every three hours would require at least 3×10^7 kw of electric power—more than the total available in any single city.

On a small scale, there are a few examples of the direct application of heat energy to change weather conditions. The warming of the air over crops to save them from frost damage has been practiced for a long time. Generally, however, the produce must have a fairly high value, such as citrus fruit has, to make the practice economically feasible. During World War II, fog over English airports was sometimes dissipated sufficiently to allow aircraft operations by burning oil to raise the air temperature locally a few degrees above the dew point. In some places, sidewalks are kept clear of snow by running hot water or steam pipes beneath the pavement. It has recently been determined that the thermal energy released to the atmosphere while heating buildings and generating power, in the densely populated areas of New York City during the winter, actually exceeds the amount received from the sun at the ground. But in these examples, we are dealing with releases of rather large amounts of heat energy over relatively small areas. These energies do affect the climatic conditions of these limited areas. But when considered over larger scales, man's input of energy into the atmosphere is, at present, minuscule. For example, the total electrical energy produced each day in the United States is equivalent to just the latent heat energy that is released during precipitation of 2.5 mm (0.1 in.) over an area of 50 km square (an area about half the size of Rhode Island).

If it seems unlikely that man has used or soon will use "brute force" to modify his atmospheric environment, except on a very small scale, how might we change the weather? It appears possible in either or both of two ways: (1) by altering the existing "natural" energy exchanges that occur among the earth's surface, the atmosphere, and space, or (2) by stimulating or "triggering" various forms of instability that sometimes arise in atmospheric processes. These two "methods" are not mutually exclusive, of course, since a change in the energy balance may trigger instabilities and the setting off of instabilities may release significant amounts of energy. An example of interference with the natural energy balance is the change that occurs when a city is built on what were green fields. (Recall the climatic differences between the city and the countryside, discussed in Chapter 7.) An example of triggering atmospheric processes is cloud seeding: Super-cooled water droplets in a cloud may be induced to freeze, grow at the expense of the remaining water drops in the cloud, and precipitate; at the same time, the release of the latent heat of fusion may provide energy for additional vertical growth of the cloud.

10-2.   Altering the Energy Exchange

Most of the unintentional changes of the environment made by man fall in this category. We recall from Chapter 3 that there are various ways by which energy is carried from one to another of the three "components" of the system—space, atmosphere, and earth's surface. Radiation is a very important "carrier" among all three; conduction transfers heat (including latent heat) between the surface and the air and, of course, convection transports heat within the air.

As we have pointed out, there are large natural variations in energy from place to place and time to time. For example, the short-wave energy absorbed at the surface varies greatly even along the same latitude because of variations in cloudiness, air purity, and surface properties. This suggests a relatively simple way to change the climate; i.e., block the radiant, conductive, and/or convective energy transport from one to another of the energy exchange components.

The amount of solar energy absorbed by the earth's surface can be altered by changing the reflectivity of either the atmosphere or the surface. There is some speculation that the planet's reflectivity or albedo has already been increased unintentionally by the fine particles with which man has increasingly polluted the air, and may be responsible for the worldwide decrease in air temperature over the past few decades. An increase in "turbidity" (the "opaqueness" of the atmosphere to solar radiation, which depends on the number and sizes of particles suspended in the air) of only 10 per cent could increase the albedo of the earth by almost 1.5 per cent, which would mean about 0.8 per cent less solar radiation to warm the earth. Such a change could lower the world's mean temperature by almost 1 °C. Although there are few systematic observations of the atmosphere's turbidity, those observations that are available indicate a 60 per cent or more increase over the last 50 years. As mentioned earlier (Chapter 1), atmospheric turbidity has been observed to have increased for 1 to 2 years following an unusually violent volcanic eruption, so it seems quite plausible that man could produce similar effects with his uninterrupted and ever-increasing pollution of the air, as illustrated by the data of Table 10-2.

The clouds in the atmosphere are the most important contributors to the overall albedo of the earth (Figure 3-9). A change of only a few per cent in the mean cloudiness of the earth would change the albedo by one or two per cent. We know that the frequency and density of

TABLE 10-2

Estimated Emissions into the Atmosphere of New York City During 1966 (Thousands of Tons)

Sources	Particles	Oxides of nitrogen	Sulfur dioxide	Hydro-carbons	Carbon monoxide
Space heating	58	115	450	14	18
Incineration	32	2	2	24	46
Transportation	10	59	20	250	1691
Industry and power generation	19	108	355	3	1
Evaporation				134	
TOTAL	119	284	827	425	1756

smog and fog over cities are higher than over the surrounding country-side, probably because of the far greater numbers of condensation nuclei that are available over cities. High-flying aircraft create contrails and add moisture and particles to the high-level atmosphere. Are we inadvertently increasing the mean cloudiness and, therefore, the albedo of the planet? It is difficult to answer this question with certainty because most of the world's long-term records of cloudiness are for heavily populated areas. But it is conceivable that man could increase the albedo by producing widespread cirrus cloud cover through seeding of the upper troposphere with rockets and aircraft. Supersonic aircraft flying at great altitudes may soon be dumping many thousands of tons of water vapor into the stratosphere each year, as well as particles that serve as freezing or condensation nuclei. In this layer, ice crystals that form could persist for many days.

The radiation-absorptive properties of gases in the air also affect the energy balance of the earth. Carbon dioxide is a good absorber of infrared radiation and is a contributor to the greenhouse effect (p. 78). Although the CO_2 concentration has undoubtedly increased over cities in the past half-century, no one is very certain about what has happened to the worldwide tropospheric levels because of the enormous fluxes into the oceans and biosphere (Table 1-2). But perhaps carbon dioxide or other infrared absorbing substances might be injected into the stratosphere; here they might help to "blanket" the earth's surface, particularly in the cold regions of the world. Ozone is an excellent absorber of the sun's ultraviolet energy. In the stratosphere, below about 25 km, ozone has a lifetime of a year or more. It might be possible to artificially destroy some of this O_3, thus perhaps permitting a little more ultraviolet radiation to penetrate into the lower atmosphere.

Ozone is also an ingredient of photochemical smog. Although its lifetime is very short near the earth's surface, perhaps it too could be destroyed more rapidly and thus prevent some of the "episodes of high-oxidant concentration" that afflict many of our cities, especially those in California.

The possibilities of modifying the earth's surface to affect radiative transfer are greater than in the atmosphere; first of all, because the surface is more accessible and, second, because the atmosphere rather quickly diffuses and purges itself of many contaminants that are injected into it. The strongly differing thermal and radiative properties of water, ice, and soil surfaces suggest ways of climatic control by altering their distribution. For example, water is sometimes used to control the temperature over crops: On a clear cold night, the temperature over an irrigated field will be a few degrees higher than over dry soil.

On a large scale, there have been several proposals for changing the climate. In the Arctic, much of the ocean is covered by ice, which effectively insulates the atmosphere from the relatively warm ocean. The mean surface temperature over the central Arctic in January is about $-30°C$ (see Figure 3-16), while in the ocean, only a couple of meters below the ice surface, the temperature is about $-2°C$. Thus, the atmosphere is separated by a short distance from a large source of heat. Not only that, but the ice pack deprives the ocean from the sunlight of summer because it reflects about 65 per cent of the insolation, whereas the water, if it were exposed, would absorb about 80 per cent. The summer sunshine would be enough to melt the ice pack if it were absorbed and not reflected. This might be accomplished by applying a thin layer of dark material, such as carbonblack, to the ice surface to reduce its high reflectivity.* Once free of ice, it is possible that the Arctic Ocean might remain uncovered even in winter. The decrease in the equator-to-pole temperature difference that this scheme would produce, if it were successful, could affect the entire global circulation and, as a consequence, worldwide climates.

A land bridge across the Bering Strait, which connects the Arctic Ocean with the Bering Sea, once existed. It has been suggested that a dam might be built across the strait, and then a greater influx of warm water from the south could be forced into the Arctic Ocean. This, too, might help to keep the Arctic Ocean free of ice.

---

* Scientists in the USSR have experimented with spreading coal dust and cinders to accelerate the melting of ice in Arctic ports and bays. They report that the spring breakup of ice can be advanced by as much as a month.

The heat exchange between other large bodies of water in the world might be controlled as, for example, that occurring between the Atlantic Ocean and the Mediterranean Sea at the Strait of Gibraltar. It is also conceivable that warm ocean currents could be diverted by barriers placed at such spots as the Florida Strait.

If the great Euro-Asian continent could be broken up somewhat more with bodies of water, perhaps the climates of the more northerly latitudes could be tempered. Perhaps large areas of lowland could be flooded, thus sharply reducing the enormous annual temperature range of northern Siberia (over 100°F), and opening up a vast area for cultivation.

10-3. Triggering Natural Instabilities

Various types of instabilities arise naturally in the atmosphere. Some of these occur in the flow of air, others in the chemical or physical properties of the atmosphere's constituents. For example, extratropical cyclones develop as the result of the growth or instability of a wave in the flow pattern. Thunderstorms and showers often develop quickly from convective instability (Figure 4-17). The suspension of cloud particles in the air is a type of "colloid system" that often becomes unstable, with some of the particles precipitating. There are also chemical reactions that go on in the atmosphere, such as those occurring in polluted air near the surface and in the upper atmosphere.

The various types of instabilities are not, of course, independent of each other. Convection can lead to phase changes and precipitation; these latter involve latent heat which, in turn, affects convection. Clouds that may be produced by convection affect the radiant energy balance and, therefore, the intensity of convection. In other words, there are no completely independent "closed" systems in the atmosphere, so that the triggering of one type of instability may induce a "chain reaction" of events that is highly complex and difficult to predict.

Cloud Seeding

It has long been known that liquid water drops frequently exist in clouds at temperatures far below 0°C. But in 1946 it was discovered that these supercooled water droplets could be converted to ice crystals by seeding with dry ice. Since the coexistence of ice crystals and water drops is one important process leading to the growth of cloud particles to precipitation size (p. 42), this discovery led to revived interest in the possibility of artificial stimulation of precipitation.

"Rainmakers" have existed in almost all cultures throughout the history of man, and cloud seeding quickly captured the public imagination. While much basic research into the physics of condensation and precipitation has taken place in the past two decades, the interest in cloud seeding has also attracted many opportunists, promising to sell rain to any farmer or community willing to pay. After more than twenty years of both scientifically planned and indiscriminate seeding, it is still difficult to obtain complete agreement on the effectiveness of seeding. The following list is a summary of the opinions and results of most reputable experimenters.

1. Seeding cannot produce condensation of water drops; clouds must already exist before seeding can have any effect.

2. Particles of solid carbon dioxide (dry ice) have the effect of momentarily cooling the droplets in their path to their freezing point; the ice crystals that are thus formed spread out by turbulent mixing to infect a large volume of the cloud.

3. It has been found that there are many chemical substances that act as "nucleating agents." Droplets that form around these substances freeze at a higher temperature than they would if they were "pure." (See p. 39.) Today, silver iodide is the most commonly used nucleating agent. Most commercial cloud seeding is done by burning an acetone solution of silver iodide in ground generators located upwind of the "target" clouds. There is often doubt as to how much of the silver iodide released from ground generators actually gets into the supercooled portions of the clouds. Also, silver iodide is photochemically sensitive; light exposure of the crystals enroute to the clouds sharply decreases their effectiveness as a nucleating agent after an hour or two.

4. Experiments in clouds have clearly demonstrated that large volumes of supercooled water droplets can be transformed into ice crystals by seeding. Large areas of ground fog and clouds composed of supercooled drops have been cleared by seeding (Figure 10-1).

5. There is as yet no very effective way of seeding "warm" clouds; i.e., those at a temperature above 0°C. Attempts have been made to initiate the coalescence process (p. 41) by intro-

FIG. 10-1. Cloud deck seeded from the air. (Note where cloud has been dissipated by seeding. Courtesy of AFCRL.)

ducing a water spray or large salt crystals into the bases of cumulus clouds.

6. Proving the effectiveness of cloud seeding in increasing the amount of precipitation requires complicated statistical analysis. This is because, at the present time, it is not possible to accurately estimate how much rain will fall naturally in a given region. Most meteorologists agree that seeding does increase precipitation somewhat under certain conditions. Convective (cumulus) type clouds behave most erratically in response to seeding, possibly because of the extremely complex pattern of turbulent motion in such clouds. Seeding appears to be most effective in increasing the precipitation from winter storms over mountains. The average increase in amount caused by seeding of precipitating clouds is 10 to 15 per cent.* There is no evidence that seeding can induce rainfall when normally there would be none. Thus, at the present time, seeding is of little value in relieving droughts.

---

* The controversy over the effectiveness of cloud seeding as a means of increasing precipitation is not ended. As this book goes to press, a report was issued on a New Hampshire study that asserts that the effect of seeding there was negative! Seeding actually reduced the precipitation amount.

Although cloud seeding has had limited success in increasing precipitation, it might be even more significant for other purposes. For example, as already mentioned, seeding has been quite effective in clearing supercooled fogs, and is now used at some airports.

There have been reports that cloud seeding has reduced the amount of hail falling in thunderstorms. Hail forms when ice crystals are carried by the violent vertical motions into the supercooled water-droplet zone of the cumulonimbus, where they grow by successive collisions with the droplets. If the number of supercooled droplets can be significantly reduced by seeding, then presumably the average size of hailstones will be smaller. Methods other than seeding have been tried to induce freezing of the supercooled water droplets in hail-producing cumulonimbus clouds. Shock waves produced by explosive charges of TNT carried into the clouds by rockets have been reported to have significantly reduced hail damage. Perhaps the annoying sonic boom of supersonic aircraft might become useful for this purpose.

Seeding might also be used to modify the "energetics" of at least some of the smaller storm systems. When supercooled water droplets are induced to freeze, latent heat of fusion is released which can help to intensify the convection within the cloud. If the convection pattern within a storm could be modified, then it is conceivable that horizontal circulation might also be changed. A few cloud-seeding experiments on hurricanes have already been carried out (by Project Stormfury under ESSA and the U.S. Navy) with this idea in mind. Although the spiral bands of clouds and precipitation extend hundreds of kilometers from the eye of a hurricane, the most intense upward air motion occurs in a relatively small ring surrounding the eye (Figure 6-10). This relatively narrow zone of great towering clouds appears to act like the "chimney" of the storm, the heat of condensation within it providing much of the "draft" that sucks the air inward at lower levels from hundreds of kilometers away. If this chimney could be widened, then the intensity of the draft and, therefore, the wind velocities might be diminished. Silver iodide, injected into the wall of clouds adjacent to the eye of the storm, may convert supercooled water droplets to ice crystals, releasing heat of fusion. The net effect might be to reduce the pressure gradient; the size of the storm would be increased, but its intensity decreased.

The first hurricane-seeding experiment was carried out in 1961, and only a few hurricanes have been seeded since then. Although some changes in clouds and the wind pattern were detected in a couple of instances, it is as yet too soon to draw any conclusions about the efficiency of the technique. But if their destructive winds could be diminished, hurricanes would be no less desirable as important producers of rain than the wave cyclones of middle latitudes.

Tornadoes are much more difficult to seed because of their small size and short duration. Recently, there has been speculation that the electrical discharges and the accompanying heating may lead to the formation of tornadoes. If electrical forces are dominant in producing tornadoes, it has been suggested that chaff (small strips of metal) dropped into the centers of high electrical charge concentration in severe thunderstorms might reduce the probability of tornado formation. Of course, even if this theory is correct, the problem still remains of how to locate these potential tornado-producers in time to be seeded.

As we indicated in the discussion on thunderstorms (p. 159), the mechanism of charge generation in clouds is not yet completely understood. Although many theories have been proposed, none seems to cover all cases that have been observed. It may be that there are several ways in which the charge may be generated, and the dominant mechanism may vary with meteorological conditions. One theory is that the charge in a cloud builds up like that in the sphere atop a Van de Graaf generator. Convection below the cloud is the "belt," rapidly carrying positive ions near the earth's surface into the heart of the cloud.

Although we are still not certain about how lightning is generated, there is a great deal of interest in learning how to suppress it. Lightning strokes damage millions of dollars worth of forests and property each year. Experimental results indicate that clouds that have been seeded by silver iodide have about a third fewer cloud-to-ground lightning strokes than those that are unseeded. No one is sure why increasing the number of ice crystals should have this effect.

As the outer layers of the atmosphere are becoming more accessible, scientists are beginning to think about ways of modifying these also. For example, chaff in the ionosphere might be used to change the concentration of electrons in this electrically conducting layer. Or, by the introduction of suitable catalysts, atomic oxygen might be induced to combine with other gases, releasing energy in the process.

10-4. Influencing the Water Cycle

Possibly the most fruitful kind of "environmental engineering" that might be undertaken is manipulation of the hydrologic cycle. Control over the water supply can be achieved in ways other than by increasing precipitation. One way is to decrease evaporation. In arid regions, the evaporation from a reservoir can be very great. For example, in the Sudan, it is estimated that the annual loss from one reservoir averages 3.1 meters. In the 17 western states of the United States, losses through evaporation from dams and reservoirs amount to about 15.6 million acre-feet (19.2×10^9 m³) each year; this is equal to all the usable water in storage in California.

Over small lakes and ponds, evaporation can be significantly reduced by reducing the speed of the wind. Thick trees can decrease the wind speed by as much as 65 per cent and the evaporation by 5 to 15 per cent. Another technique that has been used increasingly during the past decade to suppress evaporation is the spreading of a monomolecular film of a substance such as cetyl alcohol on the surface of the water. Evaporation can be retarded by 15–20 per cent in this way, but it is difficult to prevent the film from being broken by waves.

Snow is a very important natural "reservoir" of water for many places. If the rate at which snow melts could be controlled, a steady supply of water might be available throughout the year instead of having an overabundance in the spring. Increasing the rate of snow melt is not too difficult. The high reflectivity of snow can be decreased by covering the surface with some dark material such as lampblack. In Tibet, it has long been the practice to throw pebbles on snow fields to speed melting for early planting. But no practical method for retarding snow melt on a large scale has yet been suggested; slowing snow melt would be immensely valuable to regions such as California, where most of the year's water supply comes in the form of snow over the mountains.

Changes in vegetation can have significant effects on the runoff. For example, a change from trees to grass usually decreases evaporation and transpiration losses and reduces the amount of rain and snow intercepted by foliage. Vegetation changes may also change the insolation and wind speed. For example, a study in Oregon has shown that the removal of 80 per cent of the trees from a small basin caused an 85 per cent increase in streamflow during the summer, which is the dry period. When forests are removed, the surface runoff of rain is increased. Urbanization has this effect, significantly increasing the likelihood of flooding of streams.

There is also the possibility of extracting water directly from the air. Even over deserts such as the Sahara, the air typically contains about a liter of water in every 100 m³. All that is needed is sufficient cooling to extract the water. However, in some arid regions, such as along the coastline of continents and over many islands, the air is nearer to saturation than it is over the Sahara. In these places, cooling of the humid air might be accomplished by piping cold seawater from great depths offshore through condensers exposed to the air. The water obtained from the air could then be piped into reservoirs.

Along some coasts, such as that of Peru, there is little or no rain but a high incidence of fog. Here, wire screens exposed to the wind act as excellent collectors of water droplets.

10-5. Planetary Engineering

Are we moving into an age of "planetary engineering," in which man will control the earth's weather and climate? If so, the impact on the affairs of man throughout the world would be tremendous. Ridding the Arctic Ocean of its ice cover would not just affect the arctic region, as we have already pointed out. Since the atmosphere respects no boundaries, international control would be necessary.

But, at the moment, weather modification on a large scale appears to be mostly in the highly speculative, "science fiction" stage. We still have much to learn about the behavior of the atmosphere and its response to stimuli. Much further study remains on the nucleation of water particles in the atmosphere, the physics of precipitation processes, the energy exchange processes, the natural instabilities that arise in the atmosphere, and the complex chain reactions that may follow instabilities. New and better methods for exerting control over the atmosphere will undoubtedly be suggested as atmospheric sciences learn more about the details of atmospheric behavior.

Even if we do achieve a much more complete understanding of weather and climate than we now have, there will remain the problem of the effects that changes in the physical environment might have on biologic and social systems. Life as we know it is delicately tuned to the existing atmospheric environment. We must be cautious when we "tinker," lest we end with unanticipated, undesired, and perhaps, irreversible effects.

PROBLEMS

Problems marked with an asterisk (*) are the most challenging.

*1. A typical winter wave cyclone might precipitate at the rate of two inches in 24 hours. For an average-sized storm, how much latent heat is generated? Express this in calories and kilowatt-hours.

2. Compare the total energy received from the sun within the boundaries of your city on January 1 with that generated by power plants and heating units.

°3. List all of the possible chemical and physical "instabilities" within the atmosphere that are in this book or that you can think of. Can you suggest means of triggering these instabilities?

4. Perform a cloud-seeding experiment in a freezer. Cover the sides of the box with black cloth so that ice crystals will be easily seen in a beam of light. Saturate the air and disperse a few crystals of dry ice at the top of the freezer.

5. List all the ways in which you think man may be affecting the climate on various scales, both intentionally and unintentionally. Can you find any evidence from weather records that the microclimate in your area has been changed by man?

6. If you live in an area where snow falls, place a large black cloth on approximately level snow surface on a sunny day. Observe whether the snow melts faster under the covered area than elsewhere.

Appendices

A Brief Chronology of Atmospheric Science

ca. 550 B.C.	Anaximander defined wind as flowing air.
ca. 400 B.C.	Hippocrates discussed influences of climate on health in his medical treatise, "Airs, Waters, and Places."
ca. 350 B.C.	Aristotle's *Meteorologica* which, in addition to weather science, dealt with shooting stars, astronomy, and oceanography.
ca. 300 B.C.	Theophrastus' *The Book of Signs* (rules for weather forecasting).
ca. 850 A.D.	Windvanes came into use.
ca. 1330 A.D.	William Merle, at Oxford, kept first extensive weather records.
ca. 1500 A.D.	Leonardo da Vinci made improved weathervane, developed air moisture indicator.
ca. 1593	Galileo invented thermometer.
1622	Francis Bacon's treatise on winds.
1643	Torricelli invented the barometer.
1646–1648	Pascal, Periers, Descartes, demonstrated decrease of pressure with altitude.
1661	Boyle discovered gas law: pressure inversely proportional to volume.
1661	Hooke invented anemometer.
1664	Longest, continuous sequence of weather observations in the world begun at Paris.
1668	Edmund Halley, of comet fame, constructed map of the tradewinds.
1687	Dampier discovered revolving character of tropical storms.
1714	Fahrenheit temperature scale.
1735	George Hadley explained tradewinds, including effect of earth's rotation.

1736	Centigrade temperature scale.
1743	Benjamin Franklin discovered that lightning is electrical charge.
1749	Observations of temperatures above the surface using kites.
1779	Longest continuous sequence of weather observations in the United States begun at New Haven, Conn.
ca. 1780	John Black discovered carbon dioxide in the air.
ca. 1780	Joseph Priestley and Carl William Scheele discovered and isolated oxygen.
1782	Montgolfier brothers ascended in atmosphere using hot-air balloons. Hydrogen-filled balloons came into use soon after.
1783	Hair hygrometer invented by De Saussure.
1783	Antoine Lavoisier identified nitrogen and oxygen in the air.
1784	First use of meteorological balloon.
1800	John Dalton, whose chief interest was meteorology, explained variations of water vapor concentration and the role of expansion in producing condensation.
1802–1803	First cloud classification systems by Lamarck and Howard.
1804	Joseph Gay-Lussac and Jean Biot ascended to about 7 km in a balloon, took air samples.
1806	Admiral Beaufort invented wind scale for sailing.
1800–1815	First international compilation of weather observations by Chevalier de Lamarck, with Pierre La Place, Lavoisier, and others.
1821	First U.S. weather map by William Redfield.
1825	Psychrometer devised by August.
1827–1840	H. W. Dove developed atmospheric models, "Law of Storms."
1837	Poriellet invented pyrheliometer for measuring insolation.

1837	William Redfield demonstrated revolving of winds around cyclones.
1841	James Espy established rules of movement and development of storms.
1844	Gaspard de Coriolis demonstrated quantitative effects of earth's rotation on motions.
1846	M. F. Maury mapped prevailing winds and ocean currents from ship logs, thus reducing sailing times substantially.
1846	Cup anemometer invented.
1847	Vidie invented the aneroid barometer.
1855–1875	Many national meteorological services established throughout the world, public weather forecasting started. (Invention of telegraph made possible rapid dissemination of weather data.)
1865	Anticyclones discovered.
1869	First use of isobars on weather maps.
1875, 1880	Role of airborne particles as condensation nuclei demonstrated by Coulier, Aitken.
1882	Ramsay and Rayleigh identified argon in the atmosphere.
1888	H. von Helmholtz suggested importance of waves in the atmosphere.
1890–1900	Ramsay identified neon, krypton, xenon, and helium in the atmosphere.
1892–1902	Systematic use of free and captive balloons and kites carrying recording instruments began. First systematic measurements of winds in free atmosphere. First use of recording instruments.
1897	Wilson's cloud chamber demonstrated that in air free of foreign particles and ions, condensation cannot occur until relative humidity almost 800 per cent.
1902	Existence of stratosphere discovered by de Bort and Assman (from free balloon ascents of instruments). Teisserenc de Bort coined the terms "troposphere," "tropopause," and "stratosphere."
1902	Ionosphere postulated by Kennelly and Heaviside.

1913	Existence of ozone in atmosphere reported by Fabry and Buisson.
1918–1921	Theory of polar front and development of storms along it, by V. Bjerknes, H. Solberg, and J. Bjerknes.
1922	Publication of Richardson's *Weather Prediction by Numerical Process*, the first attempt at weather prediction by computation from mathematical equations.
1925–1937	Systematic observations of temperature and humidity aloft using aircraft.
1925–present	Exploration of ionosphere using radio waves.
1928–1930	Radiosonde developed, leading to regular high-level measurements of temperature, humidity, pressure, and wind.
1935	First weather radar developed in Great Britain.
1938–present	Increased density of upper-air sounding radiosonde stations permitted first detailed three-dimensional structures of the atmosphere up to about 25 km.
1940–1944	Discovery of the jet stream.
1946	Discovery by Schaefer that dry ice causes undercooled (below 0°C) water drops to crystallize. Vonnegut, in the same year, discovered that silver iodide crystals also induce these drops to freeze. These events spurred modern cloud modification and rainmaking experiments.
1946–present	Exploration of high levels of atmosphere with rockets and aircraft.
1949–present	Computation of future fields of motion in the atmosphere, with development of high-speed electronic computers.
1957	(Oct. 4) USSR launched first man-made satellite, Sputnik I.
1957–58	International Geophysical Year.
1958	(Jan. 31) United States launched its first satellite, Explorer I.
1960	(April 1) First meteorological satellite, Tiros I, launched by United States.
1962	Regular firings of high-altitude rockets begun.

Classification and Description of Clouds

A. High (base above 6 km, or 20,000 ft)
   1. Cirrus (Ci)
   2. Cirrocumulus (Cc)
   3. Cirrostratus (Cs)
B. Middle (base 2–6 km, or 6500–20,000 ft)
   4. Altocumulus (Ac)
   5. Altostratus (As)
C. Low (base below 2 km, or 6500 ft)
   6. Stratus (St)
   7. Stratocumulus (Sc)
   8. Nimbostratus (Ns)
D. Vertical development
   9. Cumulus (Cu)
   10. Cumulonimbus (Cb)

1. **Cirrus**

  Description: Delicate, white fibrous cloud, usually very thin; stars and sometimes blue sky visible through it; occasionally, parts of a halo can be seen.

  Composition: Ice crystals, normally of columnar form. (Temperature is usually below $-25°C$.)

  Formation: Occurs in air that is ascending slowly (5–10 cm s^{-1}) and steadily over a wide area; often the first cloud to appear in advance of an approaching cyclonic disturbance.

FIG. A-1.   Cirrus. (Courtesy Paul W. Nesbit.)

2. *Cirrocumulus*

Description: Thin, white cloud in the form of ripples, small waves, flakes, or globules; can be distinguished from alto-cumulus by its thinness and the small sizes of the cloud elements; stars and blue sky are visible through it; usually associated with cirrus or cirrostratus.

Composition: Ice crystals, in the forms of columns and prisms. (Temperature is usually below −25°C.)

Formation: Same as for cirrus.

FIG. A-2. Cirrocumulus and cirrus. (Note "sundog," the bright spot in lower left center.)

3. **Cirrostratus**

Description: Sheet of white or bluish cloud, sometimes slightly fibrous; sun and moon can be seen and sometimes bright stars; can be distinguished from altostratus by its much lighter appearance and occurrence of halo.

Composition: Ice crystals, in form of cubes, sometimes thick plates. (Temperature usually below $-25°$C.)

Formation: Same as for cirrus, which often precedes cirrostratus.

FIG. A-3(a).   Cirrostratus. (Courtesy Meston's Travels.)

FIG. A-3(b). Thin cirrostratus with halo. (Note upper tangent arc.)

4(a). *Altocumulus* (*undulatus*)

    Description: White, sometimes grayish clouds composed of fairly large, flattened globules that are separated by patches of sky; elements are often arranged in rows or waves. When sun or moon shines through, corona appears.

    Composition: Mostly small droplets (rarely some ice crystals in form of thick plates). (Temperature between 0° and −25°C.)

    Formation: Similar to cirrocumulus.

FIG. A-4(a). Altocumulus. (Courtesy Paul W. Nesbit.)

4(b). *Altocumulus* (*cumuliformis*)

> *Description:* Small, turreted, heaped-up clouds that form at intermediate altitudes; often predecessors of cumulonimbus and thunderstorms in the summer.
>
> *Composition:* Mostly water drops.
>
> *Formation:* Slow lifting of unstable layer of air leads to convective motion; often occur in advance of a cold front.

FIG. A-4(b). Altocumulus castellanus. (Courtesy John Dieterich, ESSA.)

4(c). *Altocumulus (lenticularis)*

    Description: Lens-shaped cloud with sharp edges; usually observed on the lee side of a mountain.

    Composition: Mostly water drops.

    Formation: Occurs in stable air when strong winds blow across mountains; clouds form in the crests of the waves that are formed downstream of the obstacle.

FIG. A-4(c). Altocumulus lenticularis. (Courtesy Paul W. Nesbit.)

5. *Altostratus*

> *Description:* Gray or bluish uniform veil having slightly stri-
> ated or fibrous structure; precipitation that usually does not
> reach the ground; sun and moon sometimes can be seen as
> though through frosted glass, and when cloud is very thin,
> a corona may be visible.
>
> *Composition:* Mixture of ice crystals and water droplets; rain-
> drops or snowflakes (depending on temperature) in bottom
> portion of cloud.
>
> *Formation:* Occurs in the air that is ascending slowly (5–10
> cm s^{-1}) over a wide area; often occurs in frontal cloud sys-
> tem between the higher cirrostratus and lower nimbo-
> stratus.

FIG. A-5. Altostratus.

6. **Stratus**

> *Description:* Resembles fog, but is above the ground; uniform, amorphous, gray layer; can be distinguished from nimbostratus by somewhat lighter shade of gray and the absence of steady rain (although light drizzle may fall); normally occurs very close to the ground (base less than 2000 ft); sun and moon not visible except when layer is very thin; when broken into ragged pieces, it is called *fractostratus*.
>
> *Composition:* Water droplets (rarely ice crystals). (Temperature usually above −5°C.)
>
> *Formation:* Lifting of a shallow, moist layer of air near the ground; radiational cooling may also frequently play a significant role.

FIG. A-6.   Stratus. (Courtesy Dudley Smith.)

7. *Stratocumulus*

> *Description:* Soft, gray clouds in the form of ridges or large globules; when clouds are close together, their undersurface has wavy appearance; sun and moon seen only through cloud edges; usually no precipitation.
>
> *Composition:* Water droplets; ice crystals rare. (Temperature usually above $-5°C$.)
>
> *Formation:* Irregular mixing over a broad area (vertical velocities up to 10 cm s^{-1}).

FIG. A-7(a). Stratocumulus.

FIG. A-7(b). Stratocumulus (vesperalis). Typically occurs as air begins to descend in the evening. (Courtesy Paul W. Nesbit.)

8. **Nimbostratus**

> *Description:* Amorphous, dark gray cloud, with base cloud to the ground; fairly steady precipitation; base usually ragged; much darker than stratus.
>
> *Composition:* Mixture of ice crystals and water droplets; snowflakes or raindrops near base, depending on temperature.
>
> *Formation:* Steady ascent (up to 20 cm s^{-1}) of air over a wide area; associated with frontal systems.

FIG. A-8. Nimbostratus. (Note virga.) (Courtesy John Dieterich, ESSA.)

9. *Cumulus*

> *Description:* Dense, vertically developed clouds with flat
> grayish bases and white, dome-shaped tops; may appear as
> isolated clouds or closely packed; sun not visible through
> cloud base. Over the oceans, base of this cloud is almost
> always near 2000 ft; over continental areas, far from large
> bodies of water, the base is much higher—typically, 3500
> to 6000 ft; the top extends as high as 20,000 ft.
>
> *Composition:* Water droplets.
>
> *Formation:* Ascent of warm air in intermittent bubbles. Up-
> ward vertical speeds to 5 ms^{-1} within cloud, 1–3 ms^{-1} be-
> low base of cloud; downdrafts, 1–2 ms^{-1}.

FIG. A-9(a). Cumulus humilis. ("Fair-weather" cumulus.) (Courtesy
J. W. Thompson.)

FIG. A-9(b). Cumulus congestus. (Courtesy Paul W. Nesbit.)

10. *Cumulonimbus*

   Description: White, dense clouds with great vertical development, having a dark base; top in form of turret, sometimes
   spreading out into an anvil shape (tops often have fibrous
   structure); precipitation in the form of rain showers that
   can be extremely heavy, sometimes with snow or hail; often
   produce thunderstorms. The cloud base is normally 2000–
   3000 feet above the ground and extends vertically to as
   high as 45,000 ft or more.

   Composition: Water droplets in lower regions, ice crystals in
   upper regions; snowflakes and hail.

   Formation: Strong convective currents. Updrafts are typically
   up to 15 ms^{-1} but sometimes exceed 30 ms^{-1} in the upper
   portion of the cloud; downdrafts can reach as much as
   25 ms^{-1} in heavy rain, but 10 ms^{-1} is more typical. May
   occur as isolated clouds or in groups along a line (squall
   line or mountain range).

FIG. A-10.   Cumulonimbus with anvil top.

Units and Conversion Tables

I. *Temperature scales*

	Fahrenheit (F)	Centigrade (C)	Kelvin (K) or Absolute (A)
Boiling point of water	212	100	373
Melting point of ice	32	0	273
Divisions between fixed points	180	100	100

Conversion formulas: $\dfrac{°F - 32}{180} = \dfrac{°C}{100}$; $°K = °C + 273°$

Fahrenheit scale to centigrade

°F	0	1	2	3	4	5	6	7	8	9
−60	−51.1	−51.7	−52.2	−52.8	−53.3	−53.9	−54.4	−55.0	−55.6	−56.1
−50	−45.6	−46.1	−46.7	−47.2	−47.8	−48.3	−48.9	−49.4	−50.0	−50.6
−40	−40.0	−40.6	−41.1	−41.7	−42.2	−42.8	−43.3	−43.9	−44.4	−45.0
−30	−34.4	−35.0	−35.6	−36.1	−36.7	−37.2	−37.8	−38.3	−38.9	−39.4
−20	−28.9	−29.4	−30.0	−30.6	−31.1	−31.7	−32.2	−32.8	−33.3	−33.9
−10	−23.3	−23.9	−24.4	−25.0	−25.6	−26.1	−26.7	−27.2	−27.8	−28.3
− 0	−17.8	−18.3	−18.9	−19.4	−20.0	−20.6	−21.1	−21.7	−22.2	−22.8
+ 0	−17.8	−17.2	−16.7	−16.1	−15.6	−15.0	−14.4	−13.9	−13.3	−12.8
10	−12.2	−11.7	−11.1	−10.6	−10.0	− 9.4	− 8.9	− 8.3	− 7.8	− 7.2
20	− 6.7	− 6.1	− 5.6	− 5.0	− 4.4	− 3.9	− 3.3	− 2.8	− 2.2	− 1.7
30	− 1.1	− 0.6	0.0	0.6	1.1	1.7	2.2	2.8	3.3	3.9
40	4.4	5.0	5.6	6.1	6.7	7.2	7.8	8.3	8.9	9.4
50	10.0	10.6	11.1	11.7	12.2	12.8	13.3	13.9	14.4	15.0
60	15.6	16.1	16.7	17.2	17.8	18.3	18.9	19.4	20.0	20.6
70	21.1	21.7	22.2	22.8	23.3	23.9	24.4	25.0	25.6	26.1
80	26.7	27.2	27.8	28.3	28.9	29.4	30.0	30.6	31.1	31.7
90	32.2	32.8	33.3	33.9	34.4	35.0	35.6	36.1	36.7	37.2
100	37.8	38.3	38.9	39.4	40.0	40.6	41.1	41.7	42.2	42.8
110	43.3	43.9	44.4	45.0	45.6	46.1	46.7	47.2	47.8	48.3

Centigrade scale to Fahrenheit

°C	0	1	2	3	4	5	6	7	8	9
−50	−58.0	−59.8	−61.6	−63.4	−65.2	−67.0	−68.8	−70.6	−72.4	−74.2
−40	−40.0	−41.8	−43.6	−45.4	−47.2	−49.0	−50.8	−52.6	−54.4	−56.2
−30	−22.0	−23.8	−25.6	−27.4	−29.2	−31.0	−32.8	−34.6	−36.4	−38.2
−20	− 4.0	− 5.8	− 7.6	− 9.4	−11.2	−13.0	−14.8	−16.6	−18.4	−20.2
−10	14.0	12.2	10.4	8.6	6.8	5.0	3.2	1.4	− 0.4	− 2.2
− 0	32.0	30.2	28.4	26.6	24.8	23.0	21.2	19.4	17.6	15.8
+ 0	32.0	33.8	35.6	37.4	39.2	41.0	42.8	44.6	46.4	48.2
10	50.0	51.8	53.6	55.4	57.2	59.0	60.8	62.6	64.4	66.2
20	68.0	69.8	71.6	73.4	75.2	77.0	78.8	80.6	82.4	84.2
30	86.0	87.8	89.6	91.4	93.2	95.0	96.8	98.6	100.4	102.2
40	104.0	105.8	107.6	109.4	111.2	113.0	114.8	116.6	118.4	120.2

II. Length

1 kilometer = 0.6214 statute mile = 0.5396 nautical mile
1 meter(m) = 1.093611 yards = 3.2808 feet = 39.370 inches
1 cm = 0.3937 in. = 10^4 microns (μ) = 10^8 angstroms (Å)

Conversion of statute miles to nautical miles and kilometers

Statute miles	Nautical miles	Kilometers
1	0.87	1.6
2	1.74	3.2
3	2.60	4.8
4	3.47	6.4
5	4.34	8.0
6	5.21	9.7
7	6.08	11.3
8	6.95	12.9
9	7.82	14.5
10	8.68	16.1
20	17.37	32.2
30	26.05	48.3
40	34.74	64.4
50	43.42	80.5
60	52.10	96.6
70	60.79	112.7
80	69.47	128.7
90	78.16	144.8

III. Velocity

1 knot (nautical mile per hour) = 1.1516 statute mph
   = 0.5148 m/sec
1 mph = 0.8684 knot = 0.447 m/sec
1 m/sec = 2.2369 miles per hour(mph) = 1.9424 knots
   = 3.2808 fps

Beaufort scale

	Beaufort force	Knots	mph	m/s
0	Calm	<1	<1	<0.4
1	Light air	1–3	1–3	0.4–1.5
2	Light breeze	4–6	4–7	1.6–3.3
3	Gentle breeze	7–10	8–12	3.4–5.4
4	Moderate breeze	11–16	13–18	5.5–7.9
5	Fresh breeze	17–21	19–24	8.0–10.7
6	Strong breeze	22–27	25–31	10.8–13.8
7	Moderate gale	28–33	32–38	13.9–17.1
8	Fresh gale	34–40	39–46	17.2–20.7
9	Strong gale	41–47	47–54	20.8–24.4
10	Whole gale	48–55	55–63	24.5–28.4
11	Storm	56–63	64–73	28.5–33.5
12	Hurricane	>63	>73	>33.5

Conversion of speeds

Knots	mph	m/sec	Knots	mph	m/sec	Knots	mph	m/sec
1	1.2	0.5	10	11.5	5.1	110	126.7	56.6
2	2.3	1.0	20	23.0	10.3	120	138.2	61.8
3	3.5	1.5	30	34.5	15.4	130	149.7	66.9
4	4.6	2.1	40	46.1	20.6	140	161.2	72.1
5	5.8	2.6	50	57.6	25.7	150	172.7	77.2
6	6.9	3.1	60	69.1	30.9	160	184.2	82.4
7	8.1	3.6	70	80.6	36.0	170	195.8	87.5
8	9.2	4.1	80	92.1	41.2	180	207.3	92.7
9	10.4	4.6	90	103.6	46.3	190	218.8	97.8
			100	115.2	51.5	200	230.3	103.0

INCHES OF MERCURY TO MILLIBARS

(Tenths and hundredths of an inch)

In. Hg	.00	.01	.02	.03	.04	.05	.06	.07	.08	.09
0.00	0.00	0.34	0.68	1.02	1.35	1.69	2.03	2.37	2.71	3.05
.10	3.39	3.73	4.06	4.40	4.74	5.08	5.42	5.76	6.10	6.43
.20	6.77	7.11	7.45	7.79	8.13	8.47	8.80	9.14	9.48	9.82
.30	10.16	10.50	10.84	11.18	11.51	11.85	12.19	12.53	12.87	13.21
.40	13.55	13.88	14.22	14.56	14.90	15.24	15.58	15.92	16.25	16.59
.50	16.93	17.27	17.61	17.95	18.29	18.63	18.96	19.30	19.64	19.98
.60	20.32	20.66	21.00	21.33	21.67	22.01	22.35	22.69	23.03	23.37
.70	23.70	24.04	24.38	24.72	25.06	25.40	25.74	26.08	26.41	26.75
.80	27.09	27.43	27.77	28.11	28.45	28.78	29.12	29.46	29.80	30.14
.90	30.48	30.82	31.15	31.49	31.83	32.17	32.51	32.85	33.19	33.53

(Tens and units of an inch)

In. Hg	0	1	2	3	4	5	6	7	8	9
0		33.86	67.73	101.59	135.46	169.32	203.18	237.05	270.91	304.78
10	338.64	372.50	406.37	440.23	474.09	507.96	541.82	575.69	609.55	643.41
20	677.28	711.14	745.01	778.87	812.73	846.60	880.46	914.33	948.19	982.05
30	1015.92	1049.78								

IV. *Force*

British: pound (lb)   1 dyne $= 2.2481 \times 10^{-6}$ pounds (lb)
Metric: dyne   1 pound $= 4.4482 \times 10^5$ dynes

V. *Pressure*

1 lb/in.$^2 = 68.947$ mb $= 2.0360$ in. Hg. $= 5.1715$ cm Hg
   $= 68,947$ dynes/cm^2

1 dyne/cm$^2 = 1.4504 \times 10^{-5}$ lb/in.$^2 = 2.9530 \times 10^{-5}$ in. Hg
   $= 7.5006 \times 10^{-5}$ cm Hg

1 mb $= 1000$ dynes/cm$^2 = 1.4504 \times 10^{-2}$ lb/in.^{2}
   $= 2.9530 \times 10^{-2}$ in. Hg $= 7.5006 \times 10^{-2}$ cm Hg

1 cm Hg $= 13,332.2$ dynes/cm$^2 = 0.19337$ lb/in.$^2 = 13.3322$ mb

1 in. Hg $= 0.49116$ lb/in.$^2 = 33,863.9$ dynes/cm$^2 = 33.8639$ mb

VI. *Energy*

1 gram-calorie [or, just "calorie," (cal)]
1 erg $= 1$ dyne cm $= 2.388 \times 10^{-8}$ cal
1 watt-hour $= 860$ gram-calories (g-cal) $= 3.600 \times 10^{10}$ ergs
1 British thermal unit (Btu) $= 0.293$ watt-hour
   $= 251.98$ gram-calories $= 1.055 \times 10^{10}$ ergs
1 joule (j) $= 10^7$ ergs
1 cal $= 4.1855 \times 10^7$ ergs
1 foot-pound $= 1.356 \times 10^7$ ergs
1 horsepower-hour $= 2.684 \times 10^{13}$ ergs $= 0.6416 \times 10^6$ cal

VII. *Power*

1 watt $= 14.3353$ cal min^{-1}
1 cal min$^{-1} = 0.06976$ watt
1 horsepower $= 746$ watts
1 Btu min$^{-1} = 175.84$ watts $= 252.08$ cal min^{-1}

Earth and Sun Data

Earth's:

Equatorial radius	6378.2 km	
Polar radius	6356.8 km	
Meridional perimeter	40,009.2 km	
Equatorial perimeter	40,075.7 km	
Surface area	510,098,073 km^2	
Volume	1,083,314 $\times$ 10^6 km^3	
Total mass	5.9737 $\times$ 10^{24} kg	
Mean density	5.517 g/cm^3	

Earth-sun distance at aphelion
($\sim$ 4 July)   152.00 $\times$ 10^6 km

Earth-sun distance at perihelion
($\sim$ 3 January)   147.01 $\times$ 10^6 km

Mean earth-sun distance   149.53 $\times$ 10^6 km

Sun's radius   695,600 km

The earth's rotation speed (angular velocity, ω, referred to in Equation 4-5) is given by:

$$\omega = \frac{2\pi}{\text{sidereal day}^*} = \frac{2\pi}{\left(\dfrac{365.25}{366.25}\right)(24 \times 60 \times 60)} = 7.292 \times 10^{-5} \text{sec}^{-1}$$

---

*The time it takes to complete one rotation relative to fixed stars. Since the earth makes one additional rotation about its axis in its annual revolution, there are 366.25 *sidereal* days compared to 365.25 *solar* days in a year. Thus, a sidereal day = 365.25/366.25 $\times$ 24 hours = 23 hr 56 min 4.09 sec.

Plotting Models and Analytic Symbols Used for Sea Level and Upper-Air Charts

PLOTTING MODEL FOR
SEA LEVEL WEATHER CHART

WW Symbols

∞	haze
≡	fog
⟩	drizzle
•	rain
✳	snow
⚇̇	rain shower
℞	thunderstorm
⚹̇	snow shower

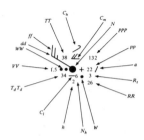

Symbols	*Example*
N: Amount of total sky cover	Overcast
ff: Barbs show wind speed (full barb = 10 knots)	25 knots
dd: Arrow shaft shows wind direction	Northwest
TT: Temperature (°F)	38°F
VV: Visibility (miles)	1.5 miles
WW: Weather type	Continuous light rain
T_dT_d: Dew-point temperature (°F)	34°F
C_l: Type of low clouds	Stratus
h: Height of ceiling	300–599 ft
N_h: Amount of low cloud cover	6/8
RR: Precipitation amount, past 6 hours	0.26 in.
W: Weather, past 6 hours	Rain
R_t: Time precipitation began or ended	Began 3–4 hours ago
a: Trend of barograph curve, past 3 hours	Rising
pp: Pressure change, past 3 hours	+2.2 mb
PPP: Sea level pressure, with only last three digits (including tenths) given	1013.2 mb
C_m: Type of middle clouds	Nimbostratus
C_h: Type of high clouds	Cirrus

PLOTTING MODEL FOR UPPER-AIR CHART

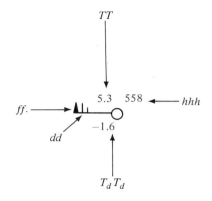

Symbols	*Example*
dd: Arrow shaft shows wind direction	270°
ff: Barbs show wind speed (triangle, 50 knots; full barb, 10 knots)	65 knots
TT: Temperature (°C)	5.3°C
T_dT_d: Dew-point temperature (°C)	−1.6°C
hhh: Height of pressure surface (in meters) with only last three digits given	1558 m (850-mb surface)

SYMBOLS FOR WEATHER CHART ANALYSIS

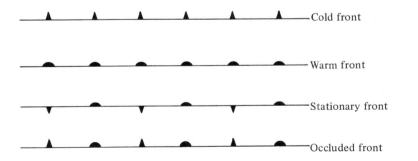

H = High pressure or height
L = Low pressure or height

Pips along fronts point in direction of movement

Aviation Weather
Reports and Forecasts

Aviation weather reports. Reports of weather conditions at airports and other locations are transmitted in the United States over special aviation teletype circuits in coded form. The following is a sample of an aviation weather report and an abbreviated explanation of the code.

Example of aviation weather report

Station and type of report	Sky and ceiling	Visibility Weather & obstruction to vision	Sea level pressure Temperature Dew point Wind
MKC	15 ① M25 ⊕	4R-K	132/58/56/1807

Altimeter setting	Runway visual range	Pilot report
/993/	R04L VR20V40	/ ⊕ 55

Explanation of aviation weather report (abridged)

Sky and ceiling
Sky cover symbols are in ascending order. Figures preceding symbols are heights in hundreds of feet above station.
Sky cover symbols are:

    ◯ Clear: less than 0.1 sky cover
    ① Scattered: 0.1 to less than 0.6 sky cover
    ⊕ Broken: 0.6 to 0.9 sky cover
    ⊕ Overcast: more than 0.9 sky cover
    − Thin (when prefixed to the above symbols)
  −× Partial obstruction: 0.1 to less than 1.0 sky hidden by pre-
       cipitation or obstruction to vision (bases at surface)
   × Obstruction: 1.0 sky hidden by precipitation or obstruction
      to vision (bases at surface)

363

Letter preceding height of layer identifies ceiling layer and indicates how ceiling height was obtained:

A   Aircraft
B   Balloon (pilot or ceiling)
E   Estimated height of non-
      cirriform clouds
M   Measured

R   Radiosonde balloon or radar
W   Indefinite
V   Immediately following nu-
      merical value, indicates a
      varying ceiling

Visibility

Reported in statute miles and fractions. (V = variable)

Weather and obstruction to vision symbols

A   Hail
BD Blowing dust
BN Blowing sand
BS Blowing snow
D   Dust
F   Fog
GF Ground fog
H   Haze

IC   Ice crystals
IF   Ice fog
IP   Ice pellets
IPW Ice pellet showers
K   Smoke
L   Drizzle
R   Rain
RW   Rain showers

S   Snow
SG   Snow grains
SP   Snow pellets
SW   Snow showers
T   Thunderstorm
ZL   Freezing drizzle
ZR   Freezing rain

Precipitation intensities are indicated thus:
$--$ Very light; $-$ Light; (no sign) Moderate; $+$ Heavy

Wind

Direction in tens of degrees from true north. Speed in knots. 0000 indicates calm. G indicates gusty. Peak speed of gusts follows G; or Q where squall is reported. The contraction WSHFT followed by local time group in remarks indicates windshift and its time of occurrence. (Kt × 1.15 = statute mph)
Examples: 3627 = 360 deg, 27 knots.
            3627G40 = 360 deg, 27 knots, peak speed in gusts 40 knots.

Altimeter setting
The first figure of the actual altimeter setting is always omitted from the report.

Runway visual range (RVR)
RVR is reported from some stations. Extreme values for 10 minutes prior to observation are given in hundreds of feet. Runway identification precedes RVR report.

Coded pireps

Pilot reports of clouds not visible from ground are coded with MSL height data preceding and/or following sky cover symbol to indicate cloud bases and/or tops, respectively.

Decoded report

Kansas City: Record observation, 1500 ft scattered clouds, measured ceiling 2500 ft overcast, visibility 4 mi, light rain, smoke, sea level pressure 1013.2 mb, temperature 58°F, dew point 56°F, wind 180°, 7 knots, altimeter setting 29.93 in. Runway 04 left, visual range 2000 ft variable to 4000. Pilot reports top of overcast 5500 ft.

Type of report

The omission of type-of-report data identifies a scheduled record observation for the hour specified in the sequence heading; the time of an out-of-sequence special observation is given as "S" followed by a time group (24-hour clock GMT); e.g., "PIT S 715-XM . . ." A special report indicates a significant change in one or more elements. Local reports are identified by "LCL" and a time group; they are transmitted on local teletypewriter circuits only.

Aviation weather forecasts. Predictions of weather conditions for aviation purposes are tailored to fit the needs of pilots for takeoff, inflight, and landing operations. These are made available throughout the United States over special aviation teletype circuits in a combination of coded symbols and plain language abbreviations. The following are (a) a simplified example of a weather forecast for an aviation terminal (airport); (b) an example of a forecast covering a large area for in-flight purposes; and (c) an example of a forecast of wind directions and speeds at various levels for a given location.

Example of a terminal weather forecast

    FT1   051045
    11Z-23Z
    OAK 35 ①   2F 1810 0430P 30 ⊕ 1F 2710

Translation of the above forecast

FT1	Terminal forecast valid for 12 hrs
05	Day of month: 5th
1045	Time forecast issued: 1045 Greenwich Mean Time (GMT)
11Z-23Z	Forecast period: 1100 GMT to 2300 GMT
OAK	Airport to which forecast applies: Oakland, Calif.
35 ①	Scattered clouds at 3500 ft
2F	Visibility 2 mi due to fog

1810	Wind from south (180 deg) at 10 knots
0430P	Weather will change at 0430 Pacific Standard Time to values which follow
30 ⊕	Overcast at 3000 ft
1F	Visibility 1 mi due to fog
2710	Wind from west (270 deg) at 10 knots

Example of area weather forecast
FA SFO 071245
05P-17P WED
NRN CALIF WRN NEV
CLDS AND WX ◯ to ◍ . TOPS 80 OVR MTNS
ICG LGT RIME ICGIC. FRZG LVL 65
TURBC MDT OVR AND ALG E SLPS OF MTNS
OTLK. 17P WED-05P THU. ◯ OVR AREA

Translation of above forecast

FA	Area forecast
SFO	Office from which issued: San Francisco, Calif.
07	Day of month: 7th
1245	Time forecast issued: 1245 Greenwich Mean Time (GMT)
05P-17P WED	Forecast period: 0500 to 1700 Pacific Standard Time, Wednesday
NRN CALIF WRN NEV	Forecast area: Northern California and Western Nevada
CLDS AND WX, etc.	Clouds and weather, clear to scattered clouds and tops of clouds 8000 ft over mountains
ICG, etc.	Icing. Light rime icing in clouds. Freezing level 6500 ft.
TURBC, etc.	Moderate over and along east slopes of mountains
OTLK, etc.	Outlook period 1700 PST Wednesday to 0500 PST Thursday. Clear over entire area

Example of winds-aloft forecast
FD1 WBC 010550
06-12Z WED

LVL	3000	5000FT	10000FT	15000FT	20000FT	25000FT
BOS	2015	2328+06	2633-01	2936-10	3043-15	3048-21

Translation of above forecast

FD1	Winds-aloft forecast
WBC	Issued from Weather Bureau National Meteorological Center
01	Day of month: 1st
0550	Time forecast issued: 0550 Greenwich Mean Time (GMT)
06-12Z WED	Forecast period: 0600 to 1200 GMT, Wednesday
LVL	Elevation above sea level: 3000 ft, 5000 ft, etc.
BOS	Location to which forecast applies: Boston, Mass.
2015	Wind at 3000 ft, 200 deg, 15 knots (temperature not forecast)
2328+06	Wind at 5000 ft, 230 deg, 28 knots; temperature +06°C
2633-01	Wind at 10,000 ft, 260 deg, 33 knots; temperature −1°C
	Other levels translated in a similar manner

Psychrometric Tables

TABLE A

Dew-Point Temperature (°F) *and Saturation Vapor Pressure* (in. Hg)
(Pressure = 30 in. Hg)

Air temperature (°F)	Vapor pressure (in. Hg)	Depression of wet-bulb thermometer (°F)														
		1	2	3	4	5	6	7	8	9	10	15	20	25	30	35
20	0.110	16	12	8	2	− 7	−21									
25	0.135	22	19	15	10	5	− 3	−15	−51							
30	0.166	27	25	21	18	14	8	2	− 7	−25						
35	0.203	33	30	28	25	21	17	13	7	0	−11					
40	0.248	38	35	33	30	28	25	21	18	13	7					
45	0.300	43	41	38	36	34	31	28	25	22	18					
50	0.362	48	46	44	42	40	37	34	32	29	26	0				
55	0.436	53	51	50	48	45	43	41	38	36	33	15				
60	0.522	58	57	55	53	51	49	47	45	43	40	25	− 8			
65	0.622	63	62	60	59	57	55	53	51	49	47	34	14			
70	0.739	69	67	65	64	62	61	59	57	55	53	42	26	−11		
75	0.875	74	72	71	69	68	66	64	63	61	59	49	36	15		
80	1.032	79	77	76	74	73	72	70	68	67	65	56	44	28	− 7	
85	1.214	84	82	81	80	78	77	75	74	72	71	62	52	39	19	
90	1.422	89	87	86	85	83	82	81	79	78	76	69	59	48	32	1
95	1.661	94	93	91	90	89	87	86	85	83	82	74	66	56	43	24
100	1.933	99	98	96	95	94	93	91	90	89	87	80	72	63	52	37
105	2.244	104	103	101	100	99	98	96	95	94	93	86	78	70	61	48
110	2.597	109	108	106	105	104	103	102	100	99	98	91	84	77	68	57
115	2.996	114	113	112	110	109	108	107	106	104	103	97	90	83	75	65

TABLE B
Relative Humidity (per cent)
(Pressure = 30 in. Hg)

Air temperature (°F)	Depression of wet-bulb thermometer (°F)														
	1	2	3	4	5	6	7	8	9	10	15	20	25	30	35
20	85	70	55	40	26	12									
25	87	74	62	49	37	25	13	1							
30	89	78	67	56	46	36	26	16	6						
35	91	81	72	63	54	45	36	29	19	10					
40	92	83	75	68	60	52	45	37	29	22					
45	93	86	78	71	64	57	51	44	38	31					
50	93	87	80	74	67	61	55	49	43	38	10				
55	94	88	82	76	70	65	59	54	49	43	19				
60	94	89	84	78	73	68	63	58	53	48	26	5			
65	95	90	85	80	75	70	66	61	56	52	31	12			
70	95	90	86	81	77	72	68	64	59	55	36	19	3		
75	96	91	86	82	78	74	70	66	62	58	40	24	9		
80	96	91	87	83	79	75	72	68	64	61	44	29	15	3	
85	96	92	88	84	80	76	73	69	66	62	46	32	20	8	
90	96	92	89	85	81	78	74	71	68	65	49	36	24	13	3
95	96	93	89	85	82	79	75	72	69	66	51	38	27	17	7
100	96	93	89	86	83	80	77	73	70	68	54	41	30	21	12
105	97	93	90	87	83	80	77	74	71	69	55	43	33	23	15
110	97	93	90	87	84	81	78	75	73	70	57	46	36	26	18
115	97	94	91	88	85	82	79	76	74	71	58	47	37	28	21

Standard Atmosphere

Altitude (m)	Temperature (°C)	Pressure (mb)	Density (kg/m³)
0	15.0	1,013.2	1.2250
500	11.8	954.6	1.1673
1,000	8.5	898.8	1.1117
1,500	5.2	845.6	1.0581
2,000	2.0	795.0	1.0066
2,500	− 1.2	746.9	0.9569
3,000	− 4.5	701.2	0.9092
3,500	− 7.7	657.8	0.8634
4,000	−11.0	616.6	0.8194
4,500	−14.2	577.5	0.7770
5,000	−17.5	540.5	0.7364
5,500	−20.7	505.4	0.6975
6,000	−24.0	472.2	0.6601
6,500	−27.2	440.8	0.6243
7,000	−30.4	411.0	0.5900
7,500	−33.7	383.0	0.5572
8,000	−36.9	356.5	0.5258
8,500	−40.2	331.5	0.4958
9,000	−43.4	308.0	0.4671
9,500	−46.7	285.8	0.4397
10,000	−49.9	265.0	0.4140
10,500	−53.1	245.4	0.3886
11,000	−56.4	227.0	0.3648
11,100	−56.5	223.5	0.3593
11,500	−56.5	209.8	0.3374
12,000	−56.5	194.0	0.3119
13,000	−56.5	165.8	0.2666
14,000	−56.5	141.7	0.2279
15,000	−56.5	121.1	0.1948
16,000	−56.5	103.5	0.1665
17,000	−56.5	88.5	0.1423
18,000	−56.5	75.6	0.1216
19,000	−56.5	64.7	0.1040
20,000	−56.5	55.3	0.0889
25,000	−51.6	25.5	0.0401
30,000	−46.6	12.0	0.0184
35,000	−36.6	5.7	0.0085
40,000	−22.8	2.9	0.0040
45,000	− 9.0	1.5	0.0020
50,000	− 2.5	0.8	0.0010
60,000	−17.4	0.225	0.000306
70,000	−53.4	0.055	0.000088
80,000	−92.5	0.010	0.000020
90,000	−92.5	0.002	0.000003

Mean Monthly and Annual Temperature (°F) and Precipitation (Inches and Tenths)

The temperature (T) and precipitation (R) data of the world are presented in seven tables: The U.S.A. (excluding Alaska and Hawaii) and six world regional lists labeled A to F. The stations in the United States list have been numbered by dividing the country into five-degree latitude belts and moving from the west to the east coast in each belt (see map, Figure A-11); thus, station number 1 is Tatoosh Island in the far northwest and the last station given is Miami in the southeast. A similar scheme has been followed in tabulating the data for the rest of the world, except that the letter prefix corresponding to that of the list has been added. (See world map, Figure A-12.)

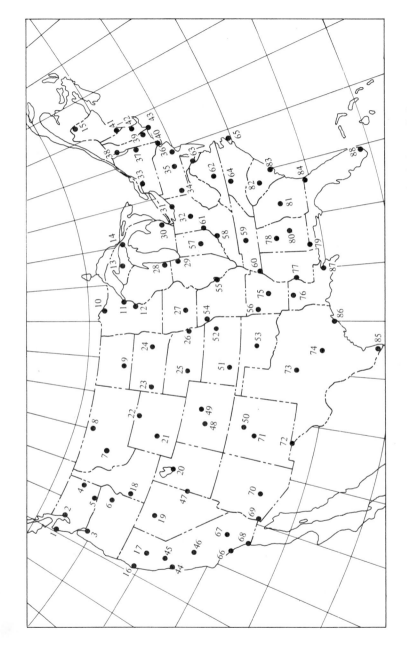

FIG. A-11. Locations of climatological stations in the United States (excluding Alaska and Hawaii).

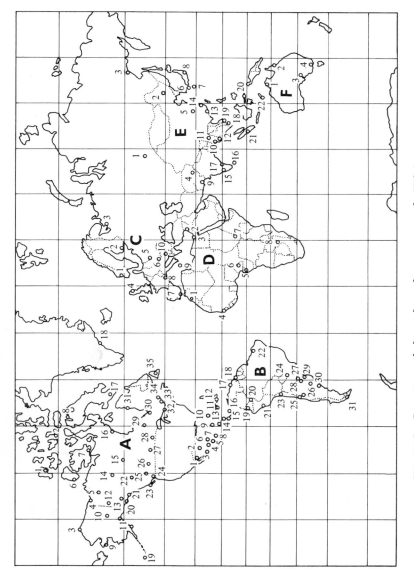

FIG. A-12. Locations of climatological stations outside U.S.A.

373

UNITED STATES (excluding Alaska and Hawaii)

		Jan.	Feb.	Mar.	Apr.	May	June	July	Aug.	Sept.	Oct.	Nov.	Dec.	Year
1. Tꜩ toosh Isl.,	T	42	43	45	47	51	54	55	55	55	52	46	45	50
Wash.	R	10.8	8.7	8.3	5.2	3.0	2.8	2.3	2.0	3.5	8.2	10.5	12.2	77.5
2. Seattle,	T	40	41	45	49	54	59	63	63	58	51	46	42	51
Wash.	R	5.7	4.2	3.8	2.4	1.7	1.6	0.8	1.0	1.8	2.1	4.0	5.4	34.5
3. Portland,	T	39	42	47	52	57	62	67	67	62	54	54	47	53
Oregon	R	5.4	4.2	3.8	2.1	2.0	1.7	0.4	0.7	1.6	3.6	5.3	6.4	37.2
4. Spokane,	T	27	30	37	48	55	61	69	68	61	48	36	30	48
Wash.	R	2.4	1.9	1.5	0.9	1.2	1.5	0.4	0.4	0.8	1.6	2.2	2.4	17.2
5. Walla Walla,	T	34	39	46	54	61	68	75	73	66	55	43	37	54
Wash.	R	1.9	1.5	1.6	1.4	1.5	1.2	0.2	0.3	0.8	1.5	1.7	1.9	15.5
6. Union,	T	30	34	39	48	54	59	66	64	57	50	39	34	46
Oregon	R	0.9	1.0	1.3	1.3	1.8	1.8	0.5	0.6	0.8	1.2	1.3	1.2	13.7
7. Helena,	T	18	23	30	41	52	59	68	66	55	45	30	25	43
Mont.	R	0.4	0.4	0.7	0.8	1.6	2.2	1.0	0.9	0.9	0.7	0.6	0.5	10.7
8. Havre,	T	16	19	28	45	55	63	72	68	57	46	32	23	45
Mont.	R	0.5	0.4	0.6	1.0	1.5	2.7	1.3	1.1	1.0	0.8	0.5	0.5	11.9
9. Bismarck,	T	8	12	25	42	55	64	70	70	59	44	28	18	40
N.D.	R	0.4	0.4	0.8	1.2	2.0	3.4	2.2	1.7	1.2	0.9	0.6	0.4	15.2
10. International	T	3	7	19	37	50	61	66	63	54	43	23	9	36
Falls, Minn.	R	0.9	0.7	1.0	1.6	2.6	3.9	3.5	3.6	2.9	1.7	1.5	0.8	24.7
11. Duluth,	T	9	10	21	31	50	59	66	64	55	45	27	14	37
Minn.	R	1.1	0.9	1.6	2.4	3.3	4.3	3.5	3.8	2.9	2.2	1.8	1.1	28.9
12. Minneapolis,	T	13	16	27	46	57	66	72	70	61	49	32	21	44
Minn.	R	0.7	0.8	1.5	1.9	3.2	4.0	3.3	3.2	2.4	1.6	1.4	0.9	24.9
13. Marquette,	T	19	19	27	39	50	61	66	66	57	48	34	25	43
Mich.	R	1.9	1.7	1.9	2.7	2.9	3.5	3.2	3.0	3.7	2.3	3.3	1.9	32.0
14. Saulte Ste.	T	13	12	22	37	49	59	64	62	56	45	32	20	39
Marie, Mich.	R	2.1	1.5	1.8	2.2	2.8	3.3	2.5	2.9	3.8	2.8	3.3	2.3	31.3
15. Presque Isl.,	T	12	14	25	39	52	61	66	64	55	45	32	18	41
Me.	R	2.2	2.1	2.3	2.5	2.8	3.8	3.9	3.3	3.3	3.4	3.0	2.4	35.0
16. Eureka,	T	48	48	49	50	54	55	57	57	57	54	52	49	52
Calif.	R	6.7	5.5	5.2	2.7	2.2	0.7	0.1	0.1	0.6	3.2	4.6	6.7	38.3
17. Red Bluff,	T	45	50	54	60	67	75	81	80	74	64	54	46	62
Calif.	R	4.4	3.6	3.0	1.7	1.0	0.5	0.1	0.1	0.6	1.3	2.8	4.3	23.4
18. Boise,	T	30	34	41	50	57	64	73	72	63	51	39	32	51
Idaho	R	1.3	1.3	1.3	1.2	1.3	0.9	0.2	0.2	0.4	0.8	1.2	1.3	11.4
19. Winnemucca,	T	29	32	37	47	54	61	71	68	59	48	36	30	48
Nev.	R	1.0	0.9	0.8	0.8	0.9	0.7	0.2	0.2	0.3	0.6	0.8	0.9	8.1
20. Salt Lake	T	29	34	41	50	59	68	76	75	64	52	37	32	52
City, Utah	R	1.4	1.2	1.6	1.8	1.4	1.0	0.6	0.9	0.5	1.2	1.3	1.2	14.1
21. Lander,	T	18	22	32	42	51	60	67	66	56	43	30	20	42
Wyo.	R	0.6	0.6	1.2	2.1	2.3	1.2	0.7	0.5	0.9	1.4	0.6	0.7	12.8
22. Sheridan,	T	21	23	30	43	54	63	72	70	59	46	34	27	45
Wyo.	R	0.6	0.7	1.4	2.2	2.6	2.6	1.2	0.9	1.2	1.1	0.8	0.6	15.9
23. Rapid City,	T	21	25	30	45	55	64	73	72	61	50	36	27	47
S.D.	R	0.4	0.5	1.0	1.7	2.7	3.1	1.8	1.2	0.9	0.8	0.4	0.3	14.8
24. North Platte,	T	25	28	34	48	59	70	77	75	64	50	36	27	50
Neb.	R	0.4	0.5	1.0	2.0	3.0	3.3	2.5	2.1	1.7	0.9	0.5	0.4	18.3
25. Huron,	T	12	16	28	45	57	68	72	73	63	48	32	19	44
S.D.	R	0.6	0.6	1.1	2.2	2.4	3.1	3.2	2.1	1.5	1.3	0.7	0.6	20.7
26. Omaha,	T	22	27	37	51	63	73	77	77	66	54	39	28	51
Neb.	R	0.7	0.9	1.5	2.5	3.5	4.5	3.5	4.0	2.6	2.2	1.3	0.8	28.0
27. Des Moines,	T	20	27	36	50	63	72	75	75	66	53	39	28	49
Iowa	R	1.3	1.1	2.1	2.5	4.1	4.7	3.1	3.7	2.9	2.1	1.8	1.1	30.5
28. Madison,	T	18	19	28	45	55	66	72	70	61	50	34	21	45
Wisc.	R	1.4	1.1	1.9	2.6	3.3	3.9	3.5	3.4	3.3	2.2	2.1	1.3	30.0
— 29. Chicago,	T	25	28	36	48	61	72	74	75	66	55	39	28	50
Ill.	R	1.9	1.6	2.8	2.8	3.7	4.1	3.3	3.1	2.7	2.5	2.2	1.9	32.6

		Jan.	Feb.	Mar.	Apr.	May	June	July	Aug.	Sept.	Oct.	Nov.	Dec.	Year
30. Detroit,	T	24	25	33	46	58	67	72	70	63	52	39	29	48
Mich.	R	2.1	2.2	2.4	2.5	3.2	3.6	3.3	2.8	2.9	2.4	2.4	2.4	32.2
31. Cleveland,	T	26	28	34	46	58	67	71	70	64	54	41	31	49
Ohio	R	2.6	2.1	2.7	2.6	2.6	2.8	3.4	2.7	3.1	2.6	2.4	2.3	31.9
32. Columbus,	T	29	32	41	51	63	73	75	75	68	55	43	34	52
Ohio	R	3.1	2.2	2.9	2.9	3.5	3.8	3.5	2.8	2.3	2.5	2.2	2.2	33.9
33. Rochester,	T	25	25	32	45	57	66	71	69	62	51	39	29	48
N.Y.	R	2.9	2.7	2.8	2.4	2.9	3.0	3.0	2.9	2.4	2.6	2.5	2.7	32.8
34. Pittsburgh,	T	31	34	41	51	63	72	75	73	68	56	45	36	53
Pa.	R	2.8	2.3	3.5	3.4	3.8	4.0	3.6	3.5	2.7	2.5	2.3	2.5	36.9
35. Harrisburg,	T	30	30	40	51	62	70	75	73	66	55	43	33	52
Pa.	R	3.0	2.7	3.1	3.0	2.9	3.7	3.5	4.1	3.4	3.0	2.0	2.9	37.3
36. Philadelphia,	T	32	34	41	52	64	72	75	75	66	55	45	34	54
Pa.	R	3.3	2.8	3.8	3.4	3.7	4.1	4.2	4.6	3.5	2.8	3.4	3.0	42.6
37. Albany,	T	24	27	36	48	61	70	73	72	64	54	41	30	50
N.Y.	R	2.5	2.2	2.9	2.9	3.6	3.7	4.3	3.3	4.0	2.8	2.9	2.7	37.8
38. Burlington,	T	19	19	28	43	55	66	70	68	61	49	37	23	45
Vt.	R	1.8	1.8	2.1	2.1	3.0	3.5	3.5	3.4	3.3	3.0	2.6	2.1	32.2
39. New Haven,	T	28	30	37	47	57	66	72	70	64	53	43	32	50
Conn.	R	3.8	3.2	4.6	3.5	3.7	3.5	4.3	4.3	3.9	3.7	4.1	4.0	46.6
40. New York,	T	31	31	38	49	60	69	74	73	69	59	44	35	52
N.Y.	R	3.3	2.8	4.0	3.4	3.7	3.3	3.7	4.4	3.9	3.1	3.4	3.3	42.3
41. Portland,	T	22	24	32	43	53	62	68	66	60	50	38	28	46
Maine	R	4.4	3.8	4.3	3.7	3.4	3.2	2.9	2.4	3.5	3.2	4.2	3.9	42.9
42. Boston,	T	28	29	36	46	57	66	72	70	63	54	42	32	50
Mass.	R	3.6	3.4	3.6	3.3	3.2	2.9	3.5	3.6	3.1	3.2	3.3	3.4	40.1
43. Block Island,	T	32	30	37	45	54	63	70	70	64	55	45	36	50
R.I.	R	3.9	3.3	4.1	3.6	3.0	2.6	2.7	3.9	3.2	3.0	3.7	3.6	40.6
44. San Francisco,	T	50	52	54	55	59	61	63	63	64	60	55	50	56
Calif.	R	4.0	3.5	2.7	1.3	0.5	0.1	0	0	0.2	0.7	1.6	4.1	18.7
45. Sacramento,	T	46	50	54	61	66	73	77	77	73	64	54	48	63
Calif.	R	3.2	3.0	2.4	1.4	0.6	0.1	0	0	0.2	0.8	1.5	3.2	16.4
46. Fresno,	T	46	51	55	61	68	75	82	80	74	65	55	47	63
Calif.	R	1.7	1.5	1.6	0.9	0.4	0.1	0	0	0.2	0.6	0.9	1.6	9.5
47. Ely,	T	23	27	34	43	50	59	68	66	59	46	34	27	45
Nevada	R	0.8	0.7	0.9	0.9	0.9	0.5	0.7	0.5	0.6	0.7	0.6	0.7	8.5
48. Montrose,	T	27	32	39	48	57	68	73	72	63	52	37	28	50
Nevada	R	0.7	0.6	0.7	1.0	0.7	0.5	0.7	1.3	0.9	0.9	0.6	0.6	9.2
49. Boulder,	T	30	34	39	47	57	68	72	73	64	51	41	36	50
Colo.	R	0.4	0.6	1.1	2.1	2.4	1.3	1.7	1.1	1.1	1.0	0.6	0.4	13.8
50. Santa Fe,	T	29	33	40	47	56	65	69	67	61	50	39	31	49
N. Mex.	R	0.7	0.8	0.8	1.0	1.3	1.1	2.4	2.3	1.4	1.2	0.7	0.7	14.4
51. Dodge City,	T	29	36	41	54	64	75	78	79	70	56	43	36	54
Kansas	R	0.5	0.7	1.1	1.9	3.2	3.0	3.1	2.4	1.5	1.3	0.6	0.5	19.8
52. Topeka,	T	28	34	41	54	64	75	81	79	70	59	43	34	55
Kansas	R	1.0	1.1	2.0	3.3	4.4	4.5	3.6	4.3	3.0	2.3	1.5	1.4	32.4
53. Oklahoma City,	T	36	41	48	60	68	77	81	81	73	61	48	39	59
Okla.	R	1.3	1.4	2.0	3.1	5.2	4.5	2.4	2.5	3.0	2.5	1.6	1.4	30.9
54. Kansas City,	T	28	31	43	55	65	74	78	77	69	58	44	32	54
Mo.	R	1.4	1.2	2.5	3.6	4.4	4.6	3.2	3.8	3.3	2.9	1.8	1.5	34.2
55. St. Louis,	T	31	34	43	56	66	77	79	77	70	59	45	36	56
Mo.	R	2.0	2.1	3.1	3.8	3.8	4.3	3.3	3.0	2.8	2.9	2.6	2.0	35.7
56. Fayetteville,	T	37	41	48	59	66	75	79	79	72	61	48	41	59
Ark.	R	2.6	3.0	3.3	4.8	6.0	5.1	3.6	3.4	4.1	3.9	3.2	2.6	45.6
57. Indianapolis,	T	29	31	41	52	63	72	76	74	67	56	42	32	53
Indiana	R	3.1	2.3	3.4	3.7	4.0	4.6	3.5	3.0	3.2	2.6	3.1	2.7	39.2
58. Louisville,	T	34	37	46	56	67	75	79	77	70	59	47	38	57
Ky.	R	4.1	3.3	4.6	3.8	3.9	4.0	3.4	3.0	2.6	2.3	3.2	3.2	41.4
59. Nashville,	T	39	43	50	59	68	77	79	79	73	61	48	41	59
Tenn.	R	5.5	4.5	5.2	3.7	3.7	3.3	3.7	2.9	2.9	2.3	3.3	4.2	45.2
60. Memphis,	T	43	45	52	62	72	79	81	81	75	63	52	45	62
Tenn.	R	4.9	4.6	4.9	5.1	4.0	3.6	3.4	2.6	2.8	2.9	4.2	4.7	47.7

United States (cont'd)

		Jan.	Feb.	Mar.	Apr.	May	June	July	Aug.	Sept.	Oct.	Nov.	Dec.	Year
61. Cincinnati,	T	33	36	43	54	64	73	78	75	70	58	45	36	56
Ohio	R	3.5	2.8	3.9	3.1	3.8	4.2	3.4	3.3	2.7	2.5	3.0	2.8	39.0
62. Lynchburg,	T	37	39	46	54	66	73	77	75	70	59	46	39	57
Va.	R	3.3	2.6	3.6	3.1	3.2	4.1	4.2	4.4	3.3	2.6	2.6	3.1	40.1
63. Washington,	T	33	37	45	53	66	73	77	77	70	57	48	37	55
D.C.	R	3.0	2.5	3.2	3.2	4.1	3.2	4.2	4.9	3.8	3.1	2.8	2.8	40.8
64. Raleigh,	T	41	43	50	59	68	76	79	77	71	62	51	43	60
N.C.	R	3.7	3.9	3.9	3.5	3.8	4.4	5.4	5.4	3.6	2.9	2.3	3.6	46.4
65. Cape Hatteras,	T	46	46	52	59	68	75	79	77	73	66	55	48	63
N.C.	R	3.9	3.9	4.2	2.3	4.0	4.1	6.1	6.4	5.9	4.3	4.1	4.6	53.8
66. Los Angeles,	T	55	57	59	59	64	68	70	73	72	65	63	59	62
Calif.	R	3.1	3.7	2.2	1.0	0.2	0.1	0	0	0.2	0.7	1.1	2.9	15.2
67. Barstow,	T	46	50	58	62	69	78	84	82	75	64	54	46	64
Calif.	R	0.8	0.6	0.7	0.2	0.1	0.1	0.2	0.2	0.2	0.4	0.3	0.6	4.4
68. San Diego,	T	55	57	59	61	63	66	67	72	70	66	61	57	61
Calif.	R	2.1	2.2	1.6	0.8	0.2	0	0	0.1	0.2	0.5	0.9	2.0	10.6
69. Yuma,	T	55	59	66	70	79	88	91	93	88	72	64	57	72
Ariz.	R	0.4	0.4	0.2	0.1	0	0	0.1	0.5	0.4	0.2	0.1	0.3	2.7
70. Phoenix,	T	51	55	63	67	77	86	90	90	84	71	59	52	70
Ariz.	R	0.8	0.9	0.7	0.4	0.1	0.1	1.1	1.1	0.7	0.5	0.5	0.9	7.8
71. Albuquerque,	T	36	39	46	55	64	75	79	77	72	59	45	37	57
N. Mex.	R	0.4	0.4	0.5	0.5	0.8	0.6	1.2	1.3	1.0	0.8	0.4	0.5	8.4
72. El Paso,	T	45	50	55	63	72	80	81	80	75	66	52	45	64
Texas	R	0.5	0.4	0.4	0.3	0.4	0.7	1.3	1.2	1.1	0.9	0.3	0.5	8.0
73. Abilene,	T	44	48	54	64	72	81	83	82	75	65	54	46	64
Texas	R	1.0	1.1	1.0	2.7	4.3	2.7	2.1	1.5	2.1	2.5	1.1	1.3	23.4
74. Austin,	T	50	54	60	68	76	83	85	82	78	71	60	52	68
Texas	R	2.4	2.6	2.1	3.6	3.7	3.2	2.2	1.9	3.4	2.8	2.1	2.5	32.5
75. Little Rock,	T	41	45	52	62	70	79	81	81	75	64	50	43	62
Ark.	R	4.7	3.9	4.8	5.2	5.3	3.6	3.5	2.8	3.2	2.7	4.1	4.1	47.9
76. Calhoun,	T	48	52	57	66	73	81	82	82	77	68	54	50	66
La.	R	5.7	4.8	4.7	4.9	5.1	3.5	4.1	2.7	2.7	2.9	4.4	5.7	51.2
77. Vicksburg,	T	47	52	57	65	73	81	82	82	77	65	55	50	65
Miss.	R	5.2	5.3	5.7	5.0	4.1	3.5	4.6	3.0	2.5	2.6	4.4	4.9	50.8
78. Birmingham,	T	46	48	55	63	72	79	82	81	77	66	54	48	64
Ala.	R	5.0	5.3	6.0	4.5	3.4	4.0	5.2	4.8	3.3	3.0	3.5	5.0	53.0
79. Mobile,	T	51	54	60	66	73	79	80	80	77	68	58	52	67
Ala.	R	4.7	5.2	6.4	4.9	4.4	5.4	7.0	7.1	5.3	3.5	3.7	4.9	62.5
80. Montgomery,	T	48	52	58	65	73	80	82	81	76	67	56	49	56
Ala.	R	5.2	5.4	6.0	4.3	3.8	3.8	4.9	4.2	3.0	2.5	3.2	4.8	51.1
81. Macon,	T	50	52	57	66	73	81	82	81	77	66	55	48	66
Ga.	R	3.4	4.3	4.9	3.7	3.3	3.3	5.6	4.2	2.8	2.0	2.4	4.0	43.9
82. Columbia,	T	46	48	54	64	72	81	82	81	75	64	54	46	64
S.C.	R	3.0	3.7	4.3	4.0	3.9	3.9	6.1	5.7	4.3	2.4	2.4	3.5	47.2
83. Charleston,	T	50	52	57	64	72	79	81	81	75	68	55	50	66
S.C.	R	3.0	3.3	3.9	2.5	3.6	5.0	6.9	6.6	5.8	3.3	2.1	2.8	48.8
84. Jacksonville,	T	55	57	63	69	75	81	82	82	79	71	63	55	69
Fla.	R	2.5	2.9	3.5	3.6	3.5	6.3	7.7	6.9	7.6	5.2	1.7	2.2	53.6
85. Brownsville,	T	59	64	68	73	79	82	84	84	81	75	68	63	73
Texas	R	1.3	1.5	1.0	1.5	2.4	3.0	1.7	2.8	5.0	3.5	1.3	1.7	26.7
86. Galveston,	T	54	57	61	70	75	82	84	84	81	73	63	57	70
Texas	R	3.4	2.9	2.9	3.1	2.8	2.6	4.0	4.4	5.1	4.3	3.5	3.9	42.9
87. New Orleans,	T	54	55	61	69	73	79	82	81	77	71	59	55	69
La.	R	3.8	4.0	5.3	4.6	4.4	4.4	6.7	5.3	5.0	2.8	3.3	4.1	53.7
88. Miami,	T	66	68	70	73	77	80	81	82	81	77	70	68	74
Fla.	R	2.0	1.9	2.3	3.9	6.4	7.4	6.8	7.0	9.5	8.2	2.8	1.7	59.9

IIA. CANADA AND ALASKA

		Jan.	Feb.	Mar.	Apr.	May	June	July	Aug.	Sept.	Oct.	Nov.	Dec.	Year
1. Mould Bay,	T	−27	−31	−26	−9	12	32	39	36	21	1	−17	−24	1
Canada	R	0.2	0.1	0.1	0.1	0.3	0.2	0.7	0.8	0.4	0.3	0.1	0.1	3.4
2. Arctic Bay,	T	−22	−26	−18	−4	19	36	43	41	28	12	−58	−17	7
Canada	R	0.3	0.2	0.3	0.2	0.3	0.4	0.7	1.0	0.9	0.7	0.3	0.2	5.5
3. Point Barrow,	T	−19	−18	−15	2	18	32	40	37	30	16	0	−11	10
Alaska	R	0.2	0.2	0.1	0.1	0.1	0.4	0.8	0.9	0.6	0.5	0.2	0.2	4.3
4. Aklavik,	T	−22	−18	−9	9	30	48	57	52	39	19	−4	−18	16
Canada	R	0.5	0.5	0.4	0.5	0.5	0.8	1.4	1.4	0.9	0.9	0.8	0.4	9.0
5. Ft. Good Hope,	T	−24	−20	−8	14	39	55	59	54	41	21	6	−20	18
Canada	R	0.7	0.6	0.6	0.5	0.7	1.2	1.9	2.2	1.3	1.2	1.1	0.6	12.6
6. Sachs Harbour,	T	−22	−26	−18	−4	16	36	43	41	28	9	−11	−18	7
Canada	R	0.1	0.1	0.2	0.1	0.2	0.2	1.0	0.7	0.7	0.4	0.2	0.2	4.1
7. Cambridge	T	−27	−31	−20	−8	16	34	46	45	32	12	−11	−22	5
Bay, Canada	R	0.3	0.2	0.2	0.2	0.2	0.5	1.0	1.0	0.6	0.6	0.4	0.2	5.4
8. Resolute,	T	−26	−29	−24	−8	14	34	41	37	23	5	−13	−20	3
Canada	R	0.1	0.1	0.2	0.2	0.5	0.8	0.9	1.1	0.8	0.5	0.2	0.1	5.5
9. Nome,	T	1	5	9	17	36	46	50	48	41	29	16	7	25
Alaska	R	1.0	0.9	0.9	0.8	0.7	0.9	2.3	3.8	2.7	1.7	1.2	1.0	17.9
10. Fairbanks,	T	−11	−2	9	30	46	59	59	55	43	27	5	−8	27
Alaska	R	0.9	0.5	0.4	0.3	0.7	1.4	1.8	2.2	1.1	0.9	0.6	0.5	11.3
11. Anchorage,	T	12	18	23	36	46	55	57	55	48	36	23	16	36
Alaska	R	0.8	0.7	0.5	0.4	0.5	1.0	1.9	2.6	2.5	1.9	1.0	0.9	14.7
12. Dawson,	T	−18	−11	5	30	46	57	59	55	43	27	3	−13	25
Canada	R	0.8	0.6	0.5	0.3	1.0	1.3	2.0	1.9	1.2	1.1	1.1	0.9	12.7
13. Whitehorse,	T	0	7	18	32	46	55	57	54	46	34	18	5	30
Canada	R	0.6	0.5	0.6	0.4	0.6	1.0	1.6	1.5	1.3	0.7	1.0	0.8	10.6
14. Fort Simpson,	T	−17	−9	5	27	46	57	63	57	46	30	7	−11	25
Canada	R	0.7	0.7	0.5	0.7	1.4	1.5	2.0	1.5	1.3	1.1	0.9	0.8	13.1
15. Fort Smith,	T	−13	−8	7	27	46	55	61	57	46	32	28	−6	27
Canada	R	0.6	0.7	0.7	0.7	1.0	1.2	2.1	1.4	1.7	1.2	1.0	1.1	13.4
16. Chesterfield,	T	−26	−26	−13	3	21	37	46	46	37	21	0	−15	10
Canada	R	0.3	0.4	0.4	0.7	0.5	1.0	1.8	1.7	1.5	1.2	0.8	0.6	10.9
17. Frobisher,	T	−15	−13	−6	9	28	39	46	45	37	23	7	−8	16
Canada	R	0.7	0.9	0.8	0.8	0.7	0.9	1.5	2.0	1.8	1.1	1.1	1.0	13.3
18. Ivigtut,	T	19	19	24	31	40	47	50	47	41	34	26	21	33
Greenland	R	3.3	2.6	3.4	2.5	3.5	3.2	3.1	3.7	5.9	5.7	4.6	3.1	44.6
19. Adak,	T	34	34	35	37	41	44	48	52	48	43	37	34	40
Alaska	R	6.6	5.3	5.9	4.2	4.6	3.1	2.9	4.3	5.5	6.9	7.3	7.6	64.2
20. Yakutat,	T	27	28	32	37	45	50	54	54	50	41	34	28	39
Alaska	R	10.9	8.2	8.7	7.2	8.0	5.1	8.4	10.9	16.6	19.6	16.1	12.3	132.0
21. Juneau,	T	27	27	30	41	46	52	57	54	48	43	34	28	42
Alaska	R	4.0	3.1	3.3	2.9	3.2	3.4	4.5	5.0	6.7	8.3	6.1	4.2	54.7
22. Prince George,	T	14	19	28	41	50	55	59	57	50	41	28	19	39
Canada	R	1.8	1.2	1.4	0.8	1.3	2.1	1.6	1.9	2.0	2.0	1.9	1.9	19.9
23. Port Hardy,	T	36	39	39	45	50	54	57	57	54	48	41	39	46
Canada	R	7.1	6.3	5.3	3.7	2.4	3.0	1.6	2.6	4.8	8.1	9.0	10.7	64.6
24. Vancouver,	T	36	39	43	48	55	59	64	63	57	50	43	39	50
Canada	R	8.6	5.8	5.0	3.3	2.8	2.5	1.2	1.7	3.6	5.8	8.3	8.8	57.4
25. Kamloops,	T	23	28	39	50	59	64	70	68	59	48	36	28	48
Canada	R	1.1	0.7	0.4	0.4	0.8	1.5	0.9	1.0	0.7	0.8	0.7	1.0	10.0
26. Calgary,	T	14	18	23	39	48	55	63	59	52	43	28	21	39
Canada	R	0.5	0.5	0.8	1.0	2.3	3.1	2.5	2.3	1.5	0.7	0.7	0.6	16.7
27. Regina,	T	1	5	18	37	52	59	66	64	54	41	23	10	36
Canada	R	0.5	0.3	0.7	0.7	1.8	3.3	2.4	1.8	1.3	0.9	0.6	0.4	14.7
28. Winnipeg,	T	0	3	18	37	52	63	68	66	55	43	23	9	37
Canada	R	0.9	0.9	1.2	1.4	2.3	3.1	3.1	2.5	2.3	1.5	1.1	0.9	21.2
29. Trout Lake,	T	−11	−6	7	25	39	52	61	59	48	34	16	−4	27
Canada	R	0.9	0.8	0.6	1.0	1.9	2.8	3.9	3.7	3.1	2.0	2.0	1.2	23.9

Canada and Alaska (cont'd)

		Jan.	Feb.	Mar.	Apr.	May	June	July	Aug.	Sept.	Oct.	Nov.	Dec.	Year
30. Moosonee,	T	—4	0	12	28	41	54	61	59	50	41	25	5	30
Canada	R	1.9	2.0	1.8	1.8	3.1	3.7	3.4	3.1	3.3	2.8	3.1	2.2	32.2
31. Port Harrison,	T	—13	—13	—4	12	28	39	48	48	41	32	18	0	19
Canada	R	0.6	0.4	0.6	0.7	0.9	1.2	2.0	2.1	2.4	1.9	1.9	0.9	15.6
32. Toronto,	T	25	25	32	45	55	66	72	70	61	52	39	28	48
Canada	R	2.7	2.4	2.6	2.5	2.9	2.7	3.0	2.7	2.9	2.4	2.8	2.6	32.2
33. Montreal,	T	16	18	28	43	57	66	72	70	61	48	36	43	45
Canada	R	3.8	3.0	3.5	2.6	3.1	3.4	3.7	3.5	3.7	3.4	3.5	3.6	40.8
34. Quebec,	T	12	13	25	38	52	62	68	65	57	45	32	18	41
Canada	R	3.5	2.7	3.0	2.4	3.1	3.7	4.0	4.0	3.6	3.4	3.2	3.2	39.9
35. Gander,	T	21	21	25	34	45	54	63	61	54	43	36	25	39
Canada	R	3.3	3.5	3.2	2.7	2.3	3.0	3.3	4.0	3.4	3.8	4.2	3.4	40.1

IIB.   MEXICO, CENTRAL and SOUTH AMERICA and THE CARIBBEAN

		Jan.	Feb.	Mar.	Apr.	May	June	July	Aug.	Sept.	Oct.	Nov.	Dec.	Year
1. Guaymas,	T	66	67	70	74	79	87	89	88	88	83	73	67	78
Mexico	R	0.6	0.3	0.3	0.1	T	T	1.5	3.5	1.3	0.5	0.3	0.8	9.2
2. Chihuahua,	T	51	54	60	67	74	82	78	77	74	66	56	49	66
Mexico	R	0.2	0.1	0.1	0.1	0.4	0.9	2.7	2.9	1.2	0.5	0.2	0.3	9.6
3. Mazatlan,	T	68	67	68	71	76	80	82	82	82	81	75	70	75
Mexico	R	0.5	0.3	0.1	0	T	1.3	6.8	8.5	9.8	2.5	0.7	1.1	31.6
4. Guadalajara,	T	59	62	67	71	73	72	69	69	68	67	64	60	67
Mexico	R	0.2	T	0.1	0.3	1.3	5.1	11.4	8.1	5.5	2.3	0.4	0.5	35.2
5. Mexico City,	T	54	57	61	64	65	64	62	62	61	59	56	54	60
Mexico	R	0.2	0.2	0.5	0.8	1.9	3.9	4.5	4.6	3.9	1.6	0.5	0.2	22.8
6. Tampico,	T	66	68	72	76	80	82	82	83	80	78	72	67	76
Mexico	R	1.5	0.7	0.5	0.7	1.9	5.6	5.9	5.1	11.7	5.7	1.9	1.2	42.4
7. Vera Cruz,	T	70	72	74	78	81	82	82	82	81	79	75	72	77
Mexico	R	0.9	0.6	0.6	0.7	2.6	11.1	14.1	11.1	13.9	6.9	3.0	1.0	66.5
8. Salina Cruz,	T	78	78	81	83	85	82	83	83	81	80	80	81	81
Mexico	R	0.2	0.1	T	0.1	2.2	10.5	6.5	6.0	9.7	3.7	0.3	0.1	39.4
9. Merida,	T	73	75	78	81	82	82	81	81	81	79	76	74	78
Mexico	R	1.2	0.9	0.7	0.8	3.2	5.6	5.2	5.6	6.8	3.8	1.3	1.3	36.4
10. Havana,	T	70	71	73	76	78	79	80	81	80	78	75	71	76
Cuba	R	2.0	1.4	2.1	2.1	6.1	5.1	3.7	4.2	6.6	7.2	2.8	2.5	45.8
11. Kingston,	T	77	77	77	78	80	81	82	82	82	81	79	78	79
Jamaica	R	1.0	0.6	1.0	1.2	4.3	4.1	1.7	3.7	4.1	7.5	3.1	1.0	33.3
12. Port-au-Prince,	T	76	77	78	79	80	81	82	81	81	80	78	77	79
Haiti	R	1.2	2.5	3.7	6.5	9.4	4.1	2.7	5.4	7.3	6.6	3.4	1.3	54.1
13. Catacamus,	T	71	74	76	79	80	78	76	76	78	77	74	73	76
Honduras	R	1.5	0.7	0.6	0.9	4.7	8.1	9.0	6.2	7.4	6.5	2.5	1.3	49.4
14. El Recreo,	T	80	82	85	85	85	84	83	82	86	84	82	80	83
Nicaragua	R	8.6	5.3	2.8	3.0	10.6	16.0	19.7	17.8	9.0	10.2	10.3	8.7	122.0
15. San Jose,	T	66	67	68	70	70	70	69	69	70	69	68	67	69
Costa Rica	R	0.3	0.2	0.4	1.5	9.6	11.2	9.1	9.2	13.5	13.1	6.8	1.8	76.7
16. Colón,	T	80	80	80	81	81	81	80	80	80	80	80	79	80
Panama	R	3.7	1.6	1.6	4.3	12.4	13.3	16.0	14.8	12.5	15.1	20.7	11.4	127.4
17. Port-of-Spain,	T	75	75	76	78	79	78	78	78	78	78	78	76	77.3
Trinidad	R	2.7	1.5	1.8	1.8	3.6	7.9	8.8	9.6	7.4	6.6	7.0	4.7	63.4

IIB. (cont'd)

		Jan.	Feb.	Mar.	Apr.	May	June	July	Aug.	Sept.	Oct.	Nov.	Dec.	Year
18.	Georgetown,	T	79	79	80	81	81	80	81	82	83	83	82	81	81
	British Guiana	R	7.9	4.6	7.2	6.0	11.1	11.7	9.9	6.5	3.1	2.9	6.7	11.1	88.7
19.	Quito,	T	56	56	56	56	57	56	56	56	57	56	56	57	56
	Equador	R	3.9	4.4	5.6	6.9	5.4	1.7	0.8	1.2	2.7	4.4	3.8	3.1	43.9
20.	Iquitos,	T	78	78	76	77	76	74	74	76	76	77	78	78	77
	Peru	R	9.1	10.4	9.4	13.6	10.7	5.7	6.4	5.2	10.5	7.3	9.1	10.3	107.7
21.	Lima,	T	70	72	72	68	64	61	60	59	59	61	63	67	65
	Peru	R	0.1	T	T	T	0.2	0.2	0.3	0.3	0.3	0.1	0.1	T	1.6
22.	Quixeramobin,	T	84	83	82	81	80	80	80	82	83	83	84	84	82
	Brazil	R	0.7	5.0	6.6	5.0	7.0	1.7	0.7	0.6	0.4	0.6	0.7	0.6	29.6
23.	Iquique,	T	71	71	69	65	63	62	60	61	64	64	67	69	66
	Chile	R					Practically	Nil							
24.	Asuncion,	T	81	80	78	72	67	63	64	66	70	73	76	80	72
	Paraguay	R	5.5	5.1	4.3	5.2	4.6	2.7	2.2	1.5	3.1	5.5	5.9	6.2	51.8
25.	Valparaiso,	T	67	66	65	61	59	56	55	56	58	59	62	64	61
	Chile	R	0.1	0	0.3	0.6	4.1	5.9	3.9	2.9	1.3	0.4	0.2	0.2	19.9
26.	Santiago,	T	67	66	62	56	51	46	46	48	52	56	61	66	56
	Chile	R	0.1	0.1	0.2	0.5	2.5	3.3	3.0	2.2	1.2	0.5	0.3	0.2	14.2
27.	Parana,	T	77	77	73	66	59	54	54	57	61	66	72	76	66
	Argentina	R	3.1	3.1	3.9	4.9	2.6	1.2	1.2	1.6	2.4	2.8	3.7	4.5	35.0
28.	Rosario,	T	77	76	70	62	56	49	51	52	57	62	69	75	63
	Argentina	R	3.7	3.2	5.3	3.1	1.8	1.5	1.0	1.5	1.6	3.5	3.4	5.3	34.9
29.	Buenos Aires,	T	74	73	69	61	55	50	49	51	55	60	66	71	61
	Argentina	R	3.1	2.8	4.3	3.5	3.0	2.4	2.2	2.4	3.1	3.4	3.3	3.9	37.4
30.	Bahia Blanca,	T	74	72	67	60	53	47	47	49	54	59	66	71	60
	Argentina	R	1.7	2.2	2.5	2.3	1.2	0.9	1.0	1.0	1.6	2.2	2.1	1.9	20.6
31.	Punta Arenas,	T	52	51	48	44	39	36	35	37	40	44	47	50	44
	Chile	R	1.5	0.9	1.3	1.4	1.3	1.6	1.1	1.2	0.9	1.1	0.7	1.4	14.4

IIC. EUROPE

			Jan.	Feb.	Mar.	Apr.	May	June	July	Aug.	Sept.	Oct.	Nov.	Dec.	Year
1.	Bergen,	T	34	34	36	42	49	55	58	57	52	45	39	36	45
	Norway	R	7.9	6.0	5.4	4.4	3.9	4.2	5.2	7.3	9.2	9.2	8.0	8.1	78.8
2.	Helsingfors,	T	21	20	25	34	46	57	62	60	52	42	32	25	40
	Finland	R	1.8	1.4	1.4	1.4	1.8	1.8	2.2	2.9	2.5	2.6	2.5	2.4	24.7
3.	Archangel,	T	8	9	18	30	41	53	60	56	46	34	22	12	33
	USSR	R	1.2	1.1	1.1	0.7	1.3	1.9	2.6	2.7	2.2	1.7	1.6	1.3	19.6
4.	Aberdeen,	T	45	45	47	50	55	60	65	64	61	56	50	46	54
	Scotland	R	2.6	2.4	2.2	2.1	2.4	1.5	1.3	1.9	2.5	3.4	3.1	3.7	29.1
5.	Breslau,	T	30	32	37	46	56	63	66	64	58	48	38	32	47
	Poland	R	1.5	1.1	1.5	1.7	2.4	2.4	3.4	2.7	1.8	1.7	1.5	1.5	23.2
6.	Spitsbergen,	T	4	−2	−2	8	23	35	42	40	32	22	11	6	18
	Norway	R	1.4	1.3	1.1	0.9	0.5	0.4	0.6	0.9	1.0	1.2	1.0	1.5	11.8
7.	Gibraltar	T	55	56	57	61	65	70	73	75	72	67	60	56	64
		R	4.6	3.4	3.7	2.5	1.4	0.2	0	0.1	0.8	3.5	4.1	5.4	29.7
8.	Marseilles,	T	44	46	50	55	61	68	72	71	66	59	51	46	57
	France	R	1.9	1.5	1.8	2.0	1.9	1.0	0.6	0.9	2.6	3.7	3.1	2.2	23.2
9.	Palermo,	T	51	52	55	58	64	71	76	77	73	67	59	53	63
	Sicily	R	3.8	3.4	2.4	1.9	1.1	0.6	0.2	0.6	2.0	3.7	4.1	4.5	28.3
10.	Belgrade,	T	29	34	43	52	62	67	72	71	63	55	43	34	52
	Yugoslavia	R	1.6	1.3	1.6	2.2	2.6	2.8	1.9	2.5	1.7	2.7	1.8	1.9	24.6

IID. AFRICA

		Jan.	Feb.	Mar.	Apr.	May	June	July	Aug.	Sept.	Oct.	Nov.	Dec.	Year
1. Marrakech,	T	52	55	59	67	69	77	82	85	76	70	62	54	67
Morocco	R	1.3	1.2	1.4	1.1	0.7	0.3	0.2	0	0.3	0.5	1.5	0.9	9.4
2. Jerusalem,	T	44	48	51	59	66	70	73	73	71	67	56	49	61
Israel-Jordan	R	5.1	4.7	2.9	0.9	0.1	0	0	0	0	0.3	2.2	3.5	19.7
3. Cairo,	T	55	57	63	70	76	80	82	82	78	74	65	58	70
Egypt	R	0.2	0.2	0.2	0.1	0.1	0	0	0	0	0	0.1	0.2	1.1
4. Bathurst,	T	74	75	76	76	77	80	80	79	80	81	79	75	78
Gambia	R	0.1	0.1	0	0	0.4	2.3	11.1	19.7	12.2	4.3	0.7	0.1	51.0
5. Douala,	T	80	80	80	79	79	77	75	75	76	76	78	79	78
Cameroun	R	1.9	3.7	8.0	8.9	12.0	21.5	29.3	27.2	20.7	16.9	6.3	2.6	159.0
6. Yaounde,	T	74	74	74	72	72	71	70	71	71	71	72	73	72.0
Cameroun	R	1.6	2.7	5.9	9.1	8.1	4.5	2.6	3.3	7.6	8.9	5.9	2.0	62.2
7. Mongalla,	T	80	82	83	81	79	77	76	76	77	78	79	79	79
Sudan	R	0.1	0.8	1.5	4.2	5.4	4.6	5.2	5.8	4.9	4.3	1.8	0.3	38.9
8. Bolobo,	T	78	79	79	78	78	78	77	78	78	77	77	77	77.9
Congo	R	5.0	7.0	4.6	7.2	5.6	0.4	0.0	2.7	3.8	6.5	7.6	10.2	60.6

IIE. ASIA

		Jan.	Feb.	Mar.	Apr.	May	June	July	Aug.	Sept.	Oct.	Nov.	Dec.	Year
1. Barnaul,	T	0	3	14	34	52	63	68	62	51	35	17	6	33
USSR	R	0.8	0.6	0.6	0.6	1.3	1.7	2.2	1.8	1.1	1.3	1.1	1.1	14.2
2. Harbin,	T	−2	5	24	42	56	66	72	69	58	40	21	3	38
Manchuria	R	0.2	0.2	0.4	0.9	1.7	3.7	6.6	4.7	2.3	1.2	0.5	0.2	22.6
3. Okhotsk,	T	−11	−7	7	21	35	45	55	55	46	27	6	−8	22
USSR	R	0.1	0.1	0.2	0.4	0.9	1.6	2.2	2.6	2.4	1.0	0.2	0.1	11.8
4. Lahore,	T	53	57	69	81	89	93	89	87	85	76	63	55	75
Pakistan	R	0.9	1.0	0.8	0.5	0.7	1.4	5.1	4.7	2.3	0.3	0.1	0.4	18.2
5. Chung King,	T	46	48	56	66	73	81	84	85	76	66	58	49	66
China	R	0.7	0.8	1.5	3.8	5.7	7.1	5.6	4.7	5.8	4.3	1.9	0.8	42.9
6. Nagasaki,	T	42	43	48	58	64	71	78	80	74	64	55	46	60
Japan	R	3.1	3.5	5.2	8.1	7.4	13.2	9.3	7.3	8.6	4.6	3.3	3.3	76.9
7. Kagoshima,	T	45	45	51	60	65	71	78	80	75	66	57	48	62
Japan	R	3.5	3.3	6.1	9.1	9.6	13.9	11.2	7.4	8.7	5.1	3.7	3.5	85.1
8. Tokyo,	T	37	39	44	55	62	69	76	78	71	60	51	41	56.9
Japan	R	1.9	2.9	4.2	5.3	5.8	6.5	5.6	6.0	9.2	8.2	3.8	2.2	61.6
9. Karachi,	T	65	68	75	81	85	87	84	82	82	80	74	67	78
Pakistan	R	0.5	0.5	0.4	0.2	0.1	0.9	2.9	1.5	0.5	0	0.1	0.1	7.7
10. Akyab,	T	70	73	79	83	84	82	81	81	82	82	78	72	79
Burma	R	0.1	0.2	0.5	2.0	13.7	49.4	53.7	42.5	24.6	11.6	5.0	0.6	203.9.
11. Mandalay,	T	70	75	83	90	89	87	87	86	85	83	76	71	82
Burma	R	0.1	0.1	0.2	1.2	5.8	6.3	2.7	4.1	5.4	4.3	2.0	0.4	32.6
12. Rangoon,	T	77	79	84	87	84	81	80	80	81	82	80	77	81
Burma	R	0.2	0.2	0.3	1.4	12.1	18.4	21.5	19.7	15.4	7.3	2.8	0.3	99.6
13. Hong Kong,	T	60	59	63	70	77	81	82	82	81	76	69	63	72
China	R	1.3	1.8	2.7	5.3	12.0	15.8	14.0	14.6	9.7	5.1	1.7	1.1	85.1
14. Foochow,	T	48	50	58	68	74	80	83	86	77	68	59	50	67
China	R	0.7	0.9	1.3	4.0	5.3	6.7	5.3	4.4	5.8	4.6	2.0	0.9	41.9
15. Cochin,	T	81	82	84	85	84	80	79	79	80	80	81	81	81.3
India	R	0.8	0.8	1.7	3.7	11.4	27.8	25.3	12.5	9.2	12.9	6.7	1.9	114.7
16. Colombo,	T	80	80	82	83	83	82	81	81	81	80	80	81	81.0
Ceylon	R	3.5	2.7	5.8	9.1	14.6	8.8	5.3	4.3	6.3	13.7	12.4	5.8	92.3
17. Madras,	T	76	78	81	85	90	90	88	86	85	82	79	77	83.1
India	R	1.4	0.4	0.3	0.6	1.0	1.9	3.6	4.6	4.7	12.0	14.0	5.5	50.0
18. Saigon,	T	79	81	84	86	84	82	82	82	82	81	80	79	81.7
S. Vietnam	R	0.6	0.1	0.5	1.7	8.7	13.0	12.4	10.6	13.2	10.6	4.5	2.2	78.1

IIE. (cont'd)

		Jan.	Feb.	Mar.	Apr.	May	June	July	Aug.	Sept.	Oct.	Nov.	Dec.	Year
19. Natrang,	T	75	77	79	82	83	84	84	84	82	80	78	76	80
S. Vietnam	R	2.4	1.1	0.9	0.9	2.4	2.2	2.0	1.5	6.9	10.6	13.9	9.6	54.4
20. Menado,	T	77	77	78	78	79	79	79	80	80	79	79	78	79
Celebes	R	18.6	14.4	10.3	8.0	6.6	6.5	4.9	3.8	3.4	4.8	8.6	14.7	104.6
21. Padang,	T	79	80	79	80	80	79	79	79	79	79	79	79	79
Sumatra	R	13.5	9.9	11.9	14.0	12.6	13.0	11.8	13.7	16.1	20.0	20.7	19.4	176.6
22. Kupang,	T	79	79	79	79	79	78	77	78	79	80	81	81	79
Timor	R	15.7	14.8	8.7	2.5	1.2	0.4	0.2	0.1	0	0.8	3.4	10.0	57.8

IIF. AUSTRALIA

		Jan.	Feb.	Mar.	Apr.	May	June	July	Aug.	Sept.	Oct.	Nov.	Dec.	Year
1. Darwin,	T	84	83	84	84	82	79	77	79	83	85	86	85	83
Northern Ter.	R	15.9	12.9	10.1	4.1	0.7	0.1	0.1	0.1	0.5	2.2	4.8	10.3	61.8
2. Cairns,	T	82	81	80	77	74	71	70	70	73	76	79	81	76
Queensland	R	30.9	22.2	32.2	22.2	13.2	8.0	4.2	5.4	3.7	3.8	8.1	11.7	165.6
3. Adelaide,	T	74	74	70	64	58	54	52	54	57	62	67	71	63
So. Australia	R	0.8	0.7	1.0	1.8	2.7	3.0	2.6	2.6	2.1	1.7	1.1	1.0	21.1
4. Melbourne,	T	68	68	65	60	54	51	49	51	54	58	61	65	59
Victoria	R	1.9	1.8	2.2	2.3	2.1	2.1	1.9	1.9	2.3	2.6	2.3	2.3	25.7

U. S. Meteorological Satellites

TIROS	I	II	III	IV	V	VI	VII	VIII	IX	X
Launch date	4/1/60	11/23/60	7/12/61	2/8/62	6/19/62	9/18/62	6/19/63	12/21/63	1/22/65	7/1/65
End of operations	6/19/60	2/1/61	10/30/61	6/12/62	5/5/63	10/11/63	2/3/66	1/22/66	2/15/67	6/1/66
Apogee (km)	740	727	821	843	971	711	649	752	2,578	838
Perigee (km)	701	625	736	706	589	684	621	701	703	752
Inclination angle° (degrees)	48	48	48	48	58	58	58	58	84	81
Orbital period (minutes)	99	98	100	100	101	99	97	100	119	101

NIMBUS	I	II
Launch date	8/24/64	5/15/66
End of operations	9/23/64	
Apogee (km)	929	1,179
Perigee (km)	423	1,112
Inclination angle° (degrees)	81	80
Orbital period (minutes)	98	108
Approximate surface area (km²) viewed by cameras and resolution (km)		4,120/3.0
Radio frequency† (mc/s)		136.95

ATS	I	II	III
Launch date	12/6/66	4/5/67	11/5/67
End of operations			
Apogee (km)	35,838	10,434	36,170
Perigee (km)	35,822	194	35,822
Inclination angle° (degrees)	0.2	28	0.5
Orbital period (minutes)	1,436	209	1,445
Approximate surface area (km²) viewed by cameras and resolution (km)	2×10^8/3.5		2.3×10^8/3.5
Radio frequency† (mc/s)	135.60		135.60

ESSA	I	II	III	IV	V	VI	VII	VIII	IX
Launch date	2/3/66	2/28/66	10/2/66	1/26/67	4/20/67	11/10/67	8/16/68	12/15/68	2/26/69
End of operations	5/8/67		10/9/68						
Apogee (km)	840	1,416	1,486	1,440	1,423	1,486	1,469	1,461	1,506
Perigee (km)	697	1,353	1,382	1,329	1,353	1,407	1,432	1,413	1,423
Inclination angle* (degrees)	82	79	79	78	78	78	79	78	79
Orbital period (minutes)	100	114	115	113	114	115	114	115	115
Approximate surface area (km²) viewed by cameras and resolution (km)		5,940/3.7		5,940/3.7		5,940/3.7	4,000/8.1	11,127,562/8.3	11,127,562/8.3
Radio frequency† (mc/s)		137.50		135.60		137.50		137.62	

* Angle between orbital and equatorial planes (highest latitude reached).
† For reception of automatic picture transmission.

Bibliography

Elementary books on general meteorology and climatology:

Battan, Louis J., *The Nature of Violent Storms*. Garden City, N.Y.: Doubleday & Co., Inc., 1961.

Bates, D. R., ed., *The Earth and its Atmosphere*. New York: Basic Books, Inc., 1957.

Blair, Thomas A. and Robert C. Fite, *Weather Elements*. Englewood Cliffs, N.J.: Prentice-Hall, Inc., 1965.

Blumenstock, David Irving, *The Ocean of Air*. New Brunswick, N.J.: Rutgers University Press, 1959.

Brooks, C. E. P., *Climate in Everyday Life*. New York: Philosophical Library, Inc., 1951.

Dobson, G. M. B., *Exploring the Atmosphere*. New York: Oxford University Press, 1963.

Dunn, Gordon E. and Banner I. Miller, *Atlantic Hurricanes*. Baton Rouge, La.: Louisiana State University Press, 1964.

Geiger, Rudolph, *The Climate Near the Ground*. Cambridge, Mass.: Harvard University Press, 1957.

Hare, F. K., *The Restless Atmosphere*. New York: Harper and Row Publishers, Inc., 1963.

International Cloud Atlas (complete and abridged editions). World Meteorological Organization, Geneva, Switzerland, 1956.

Lehr, Paul E., R. W. Burnett, and H. S. Zim, *Weather*. New York: Golden Press, 1957.

Mason, Basil J., *Clouds, Rain, and Rainmaking*. New York: Cambridge University Press, 1962.

Neuberger, Hans and F. B. Stephens, *Weather and Man*. Englewood Cliffs, N.J.: Prentice-Hall, Inc., 1948.

Pettersen, S., *Introduction to Meteorology*. New York: McGraw-Hill Book Co., Inc., 1958.

Riehl, Herbert, *Introduction to the Atmosphere*. New York: McGraw-Hill Book Co., Inc., 1965.

Spar, Jerome, *Earth, Sea, and Air. A Survey of the Geophysical Sciences*. Reading, Mass.: Addison-Wesley Publishing Co., 1965.

Sutton, O. G., *The Challenge of the Atmosphere*. New York: Harper & Bros., 1961.

Taylor, George F., *Elementary Meteorology*. Englewood Cliffs, N.J.: Prentice-Hall, Inc., 1954.

Trewartha, Glenn T., *An Introduction to Climate*. New York: McGraw-Hill Book Co., Inc., 1954.

U.S. Weather Bureau—Federal Aviation Agency, *Aviation Weather*. Washington, D.C., 1965.

Some sources of weather data and useful periodicals:

The Smithsonian Institution, Washington, D.C. "World Weather Records."

U.S. Weather Bureau (ESSA), Washington, D.C.:
Average Monthly Weather Resume and Outlook
Climates of the States
Climatic Charts for the United States
Climatological Data for the U.S. by Sections
Daily Weather Map
Monthly Climatic Data for the World

Weather, published monthly by the Royal Meteorological Society, London, England.

Weatherwise, published bimonthly by the American Meteorological Society, Boston, Mass.

Index

Index